Designed cover image: Getty images

First edition published 2026
by CRC Press
2385 NW Executive Center Drive, Suite 320, Boca Raton FL 33431

and by CRC Press
4 Park Square, Milton Park, Abingdon, Oxon, OX14 4RN

CRC Press is an imprint of Taylor & Francis Group, LLC

ISBN: 978-1-032-85700-8 (hbk)
ISBN: 978-1-032-88998-6 (pbk)
ISBN: 978-1-003-54073-1 (ebk)

DOI: 10.1201/9781003540731

Typeset in Palatino
by Apex CoVantage, LLC

Contents

About the Authors

Farzad Ebrahimi is an Associate Professor in the Department of Mechanical Engineering, IKIU, Qazvin, Iran. His research interests include mechanical behaviors of Nano-Engineered Systems, mechanics of composites and nanocomposites, functionally graded materials, viscoelasticity, and smart materials and structures. Dr. Ebrahimi has authored more than 400 high-quality peer-reviewed research articles in his fields of interest. He has also edited and authored multiple books for well-known publishers. He serves as an Associate Editor or Editorial Board member for several reputable journals, and is a distinguished reviewer whose expertise helps the editors of prestigious journals judge research articles.

Mehrdad Farajzadeh Ahari graduated as a master's student from Imam Khomeini International University as one of the top five students. His research interests are the dynamic behavior of smart materials, functionally graded materials, and shape memory materials. He has authored articles in his research field in well-known journals. Dedicated and experienced mechanical engineer with extensive knowledge of engineering principles, theories, specifications, and standards.

Hossein Sepiani earned his Doctor of Philosophy in Civil and Environmental Engineering from the University of Waterloo, after completing his Master of Science in Mechanical Engineering at the University of Tehran. His research has centered on viscoelastic and viscoplastic materials, with a strong emphasis on the development of constitutive equations, as well as advanced numerical evaluation and analysis. In addition, he has conducted extensive studies in advanced materials and structures, with his work primarily focusing on composites, polymers, smart materials, and functionally graded materials. He has published numerous peer-reviewed studies in prestigious academic journals, contributing valuable insights to the field.

Preface

The engineering and science of active materials is an active and lively frontier, one in which the ability to respond to external stimulus – temperature, light, electrical, or magnetic fields – holds open entirely new realms of possibility. Such materials, capable of converting energy to mechanical work or of changing properties in real-time, bridge static engineering solutions and the adaptive, autonomous responses within biological systems. From shape memory alloys recovering their form when heated to electroactive polymers that mimic muscle movement, active materials are revolutionizing the frontiers in robotics, biomedical devices, energy harvesting, and advanced manufacturing. Their multidisciplinary source, drawn from mechanics, physics, chemistry, and biology, sparks a fascinating research environment that can potentially reshape the future of technology and sustainability.

This book *Mechanics of Active Materials and Structures* is an exhaustive coverage of the science, the properties, and the transformative potential of active materials. It is planned to serve as a definitive reference for scientists, as well as a basic textbook for students and practitioners who want to learn about this rapidly emerging field. The chapters in this book offer a comprehensive tour of the engineering and science of active materials, beginning with an overview of their intrinsic properties and classifications (Chapter 1). Subsequent chapters address their wide-ranging applications (Chapter 2), the mechanics of their vibration and buckling behavior at the nanoscale (Chapters 3 and 4), and the advanced technologies driving their development (Chapter 5).

Each chapter is composed with a balance of theory, reality, and visionary thinking. Chapter 1 sets the context by defining active materials and their unique properties, such as self-organization and stimuli-responsiveness and also referencing key scientific milestones. Chapter 2 speaks to the versatility of application of active materials to technology fields ranging from telecommunication to medicine and stresses the solutions they can offer to solve engineering problems of a multifaceted nature. Chapters 3 and 4 give in-depth information on the vibrational and buckling properties of active material-based nanobeams and nanoplates, providing precious advice for the design of nanoscale devices. Lastly, Chapter 5 explores new technologies, from smart textiles to AI-aided materials design, which will usher in the boundaries of what active materials can achieve.

This book is aimed at a broad audience of material scientists, engineers, physicists, and technologists, as well as policymakers and industry executives who are interested in the social and economic implications of active materials. By providing fundamental principles and current advances, our hope is that it will foster additional research, enable interdisciplinary cross-talk, and propel the realization of active material technologies into tangible applications. As we stand on the threshold of the beginning of a new age in material science, we hope that this book will be a source of inspiration for innovation and encourage readers to explore the limitless potential of active materials for building a more resilient and sustainable future.

Prof. Farzad Ebrahimi
Imam Khomeini International University

Acknowledgments

The authors are eager to extend their gratitude to those who played significant roles in bringing this book to life. The authors would also like to express their gratitude to Taylor & Francis Group/CRC Press for their assistance with the proposed book. The writers significantly depended upon their wisdom and friendliness. The authors' special gratitude goes to Jonathan Plant and Bhavna Saxena, executive editor and editorial assistant at CRC Press, for their indispensable efforts during the publication process.

1

An Introduction to Active Materials

1.1 Overview of Active Materials

Active materials are a fascinating class of materials that have the ability to convert various forms of energy – chemical, thermal, electrical, or optical – into mechanical work or motion. As opposed to passive materials, which respond to stimulation through static deformation or phase transformation, active materials exhibit dynamic, self-sustained processes. These processes result from internal dissipation or generation of energy and enable applications in soft robotics, biomedical devices, energy harvesting, and adaptive structures. The defining feature of active materials is that they are capable of undertaking tasks autonomously or semi-autonomously and tend to reproduce biological systems such as muscles, cilia, or bacterial colonies [1].

Active materials consist of a wide range of systems, including liquid crystals [2], hydrogels [3–5], shape-memory alloys [6], piezoelectric materials [7], and active colloids. They consist of structures on various length scales, ranging from assemblies at the molecular level to macroscopic composites, all tailored to achieve desired functions. The topic has attracted an extreme amount of attention as it is interdisciplinary with mechanics, material science, physics, chemistry, and biology. This chapter presents the basic concepts, classifications, and structural diversity of active materials, and emphasizes the most important scientific achievements that have directed their development.

1.2 Defining Characteristics and Principles

The defining property of active materials is their ability to store energy in order to produce mechanical responses. This is typically enabled by

DOI: 10.1201/9781003540731-1

non-equilibrium processes, where the addition of energy makes the material move away from thermodynamic equilibrium, resulting in dynamic effects such as self-propulsion, oscillation, or shape morphing. Active liquid crystals, for instance, can flow spontaneously due to molecular alignment changes, while hydrogels expand or contract when they are subjected to environmental stimuli like pH [8]or temperature.

Active material mechanics are regulated by non-equilibrium thermodynamic laws and continuum mechanics. The prevailing phenomena are as follows:

- Self-Organization: Active materials tend to create ordered structures or patterns, like vortices in active fluids or traveling waves in polymer gels, due to internal energy dissipation. In Figure 1.1, mechanisms of formation for mechanically and chemically cross-linked network hydrogels are presented.
- Stimuli-Responsiveness: External stimuli (e.g., light, electric fields) tune the material's response, allowing for programmable behavior [10, 11].
- Energy Conversion: Efficient energy conversion of heat or chemical energy to mechanical work, as illustrated by shape-memory alloys reverting to preselected configurations when heated [12, 13].

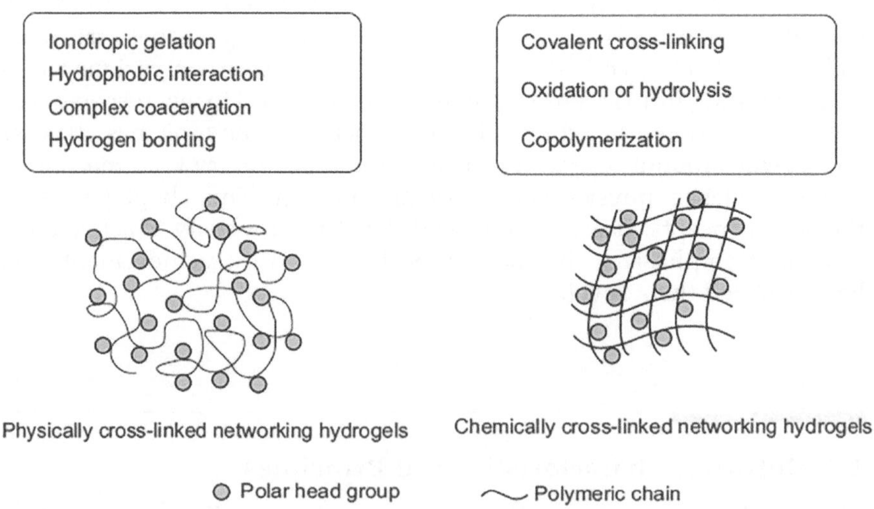

FIGURE 1.1
Mechanisms of formation for mechanically and chemically cross-linked network hydrogels [9].

These features distinguish active materials from traditional engineering materials and offer possibilities for the creation of adaptive, life-like functionality in systems.

1.3 Structural Diversity of Active Materials

Active materials are classified according to their composition, structure, and activation mechanisms. Their structural diversity is stressed in the succeeding categories:

1.3.1 Polymeric Active Materials

Polymeric networks, such as hydrogels and elastomers, are widely investigated for their elasticity and tunable properties. Hydrogels, which are cross-linked polymer networks that are water-swollen, respond to external stimuli like temperature, pH, or light by changing volume. Thermoresponsive poly(N-isopropylacrylamide) (PNIPAM) hydrogels, for example, shrink at temperatures greater than a critical temperature, and these can be used for drug delivery and soft actuator systems. Liquid crystalline elastomers (LCEs) combine the orientational order of liquid crystals and the elasticity of polymers with reversible shape changes under thermal or optical stimuli [14].

1.3.2 Metallic and Inorganic Active Materials

Shape-memory alloys (SMAs) such as Nitinol (nickel-titanium) are metal active materials that restore their original shape upon heating after deformation. Their behavior is induced by reversible martensitic phase transformation. Piezoelectric ceramics such as lead zirconate titanate (PZT) exhibit mechanical strain in electric fields and find application in sensors and actuators. They are highly sought after for high stiffness and high response speed [15].

1.3.3 Active Colloids and Fluids

On the nanoscale and microscale, colloids such as unbalanced Janus particles exhibit self-propulsion from chemical reaction or external fields. Active fluids such as bacterial suspensions or artificial microswimmers exhibit collective behaviors such as swarming or turbulence through the activity of the individual units. These systems are models to study non-equilibrium dynamics and micro-robot design [16].

1.3.4 Bioinspired and Biological Active Materials

Nature-inspired, bioinspired active materials are meant to replicate the behavior of biological tissues. For instance, myofibril contractility is replicated by synthetic muscle-like actuators via the utilization of dielectric elastomers or carbon nanotube yarns. Active biological materials like cytoskeletal proteins (myosin and actin) provide us with a vision to create synthetic systems with the same self-organization [17].

Each structural class has its own unique advantages, ranging from hydrogels' softness to SMAs' strength, to enable tailored solutions for a wide range of applications.

1.4 Scientific Breakthroughs in Active Materials Research

Active material research has been spurred by groundbreaking scientific work that has spanned decades. Early work on shape-memory alloys in the 1960s provided the foundation of phase transformation principles in active systems. Discovery by Buehler et al. in 1963 of Nitinol's shape-memory effect and superelasticity paved the way for medical stent and aerospace component applications [18].

In the 1990s, Tanaka and colleagues' responsive hydrogels were promising evidence of the application of soft matter in actuation and sensing. Their studies of critical gel transitions also illuminated how molecular interactions govern macroscopic behavior. At approximately the same time, advances in liquid crystals by de Gennes and others also revealed the active, dynamic character of nematics, which then led to the applications of tunable optics and soft robotics [19].

The 2000s also saw research on active colloids and microswimmers take off, driven by the desire to explore biological motility and synthesize biological counterparts through engineering. Paxton et al.'s research on catalytic Janus particles opened the door to self-propulsion on the micro scale, opening up new possibilities for nanotechnology. Bioinspired concepts have been at the forefront in recent years, with work like that of Whitesides and collaborators on soft robotic systems demonstrating the potential for active materials to mimic biological processes [20].

Theoretical frameworks too have come a long way. Marchetti et al.'s building of active matter theory in the 2010s provided a shared vocabulary to characterize non-equilibrium systems, ranging from bacterial suspensions to artificial active gels. These theories have been guiding experiments and contributing to our comprehension of collective dynamics.

1.5 Applications and Future Directions

Active materials have transformed numerous fields. In biomedical engineering, hydrogels and SMAs enable minimally invasive devices like drug delivery systems and stents. LCEs and dielectric elastomers power soft actuators in robotics that move through difficult environments. Piezoelectric materials are utilized by energy harvesting devices to convert mechanical vibrations into electricity, while active colloids can potentially provide targeted drug delivery and environmental remediation.

Future challenges are scaling up active material systems, enhancing energy efficiency, and having exact control over dynamic properties. Future trends include adding machine learning for learning in evolving material design and creating hybrid systems combining multiple active mechanisms. Biology continues to be a wonderful source of inspiration in the field, where researchers attempt to create materials with autonomous decision-making capabilities like living things.

1.6 Conclusion

Active materials are a rapidly evolving area of research between mechanics, material science, and physics. With their ability for energy conversion to motion and their structural complexity, they have provided the basis for groundbreaking applications and stimulated vast scientific interest. From the shape memory of metal alloys to autonomic motion in colloids, active materials push beyond classical material behavior and open new paths to future technology. Providing the introductory framework for detailed examination of their mechanics, properties, and applications in the pages that follow is a token of respect for the sheer effort of researchers who assembled this vibrant discipline.

References

1. Ebrahimi, F., and M.F. Ahari, *Mechanics of Active Materials*. 2024: Elsevier.
2. Zhang, R., A. Mozaffari, and J.J. de Pablo, Autonomous materials systems from active liquid crystals. *Nature Reviews Materials*, 2021. 6(5): p. 437–453.
3. Dong, Y., A.N. Ramey-Ward, and K. Salaita, Programmable mechanically active hydrogel-based materials. *Advanced Materials*, 2021. 33(46): p. 2006600.

4. Guo, Y., et al., Hydrogels and hydrogel-derived materials for energy and water sustainability. *Chemical Reviews*, 2020. 120(15): p. 7642–7707.
5. Zhang, Y.S., and A. Khademhosseini, Advances in engineering hydrogels. *Science*, 2017. 356(6337): p. eaaf3627.
6. Jani, J.M., et al., A review of shape memory alloy research, applications and opportunities. *Materials & Design (1980–2015)*, 2014. 56: p. 1078–1113.
7. Shivashankar, P., and S. Gopalakrishnan, Review on the use of piezoelectric materials for active vibration, noise, and flow control. *Smart Materials and Structures*, 2020. 29(5): p. 053001.
8. Alapan, Y., et al., Reprogrammable shape morphing of magnetic soft machines. *Science Advances*, 2020. 6(38): p. eabc6414.
9. Nayak, A.K., and B. Das, Introduction to polymeric gels, in *Polymeric Gels*. 2018: Elsevier. p. 3–27.
10. Ahn, S.-K., et al., Stimuli-responsive polymer gels. *Soft Matter*, 2008. 4(6): p. 1151–1157.
11. Zheng, R., et al., Stimuli-responsive active materials for dynamic control of light field. *Responsive Materials*, 2023. 1(2): p. e20230017.
12. Mercier, O., and K. Melton, Theoretical and experimental efficiency of the conversion of heat into mechanical energy using shape-memory alloys. *Journal of Applied Physics*, 1981. 52(2): p. 1030–1037.
13. Schmidt, H., Magneto-mechanical energy conversion in magnetic shape memory alloys, in *Journal of Physics: Conference Series*. 2011: IOP Publishing.
14. Kim, S., K. Lee, and C. Cha, Refined control of thermoresponsive swelling/deswelling and drug release properties of poly (N-isopropylacrylamide) hydrogels using hydrophilic polymer crosslinkers. *Journal of Biomaterials Science, Polymer Edition*, 2016. 27(17): p. 1698–1711.
15. Naresh, C., P. Bose, and C. Rao, Shape memory alloys: a state of art review, in *IOP Conference Series: Materials Science and Engineering*. 2016: IOP Publishing.
16. Zöttl, A., and H. Stark, Emergent behavior in active colloids. *Journal of Physics: Condensed Matter*, 2016. 28(25): p. 253001.
17. Banerjee, S., M.L. Gardel, and U.S. Schwarz, The actin cytoskeleton as an active adaptive material. *Annual Review of Condensed Matter Physics*, 2020. 11(1): p. 421–439.
18. Balaapgol, B., S.A. Kulkarni, and K.M. Bajoria, A review on shape memory alloy structures. *International Journal of Acoustics and Vibrations*, 2004. 9(2): p. 61–68.
19. Tanaka, T., Gels. *Scientific American*, 1981. 244(1): p. 124-S-17.
20. Sadeghi, A., *Controlled Motion of Polymeric Microparticles at Interfaces and in Solution*. 2014: University of Sheffield.

2

Applications of Active Materials

2.1 Background

Generally, the active material represents one of the most promising classes of transformational materials in science and engineering, capable of changing its properties or configurations as an outcome of external stimuli. Active materials are empowered with different types of triggers such as temperature, light, magnetic, and electrical fields, and/or mechanical forces toward adaptive actuation or autonomous self-healing and realization of certain functions. These unique and dynamic materials have thus stimulated enormous activity in many areas, resulting in the realization of their uses in various forms.

The mechanisms inherent in each class of active material are responsible for such variety. For example, SMAs return to their original form when heated, whereas piezoelectric materials produce electrical charges under mechanical loads. Similarly, electrochromic materials change their optical properties with voltage applied and thus are used in the new generation of energy-efficient smart windows. In this vein, these phenomena may indicate that versatility really can give active materials much greater potential for solving more complex challenges.

The growing sophistication in technology has increased the demand for materials that could do more than just static functionalities. Active material applications have started to appear in robotics, healthcare, space, energy, and beyond. From artificial muscles to self-regulating implants, from dampers to energy-harvesting devices, these materials continue to push the edges of innovation. In this chapter, the basic principles of active materials, their classification, and their transformative applications in key sectors are discussed, underlining their potential to shape the future of technology and sustainability.

DOI: 10.1201/9781003540731-2

2.2 Utilization of Metamaterials as a Category of Active Materials

Metamaterials, with their unique and intentional features, have opened several applications in many fields. A few of the key implementations of metamaterials, apart from active materials, are given next:

As illustrated in Figure 2.1, metamaterials can be designed to produce compact, high-gain antennas that increase signal reception and transmission efficiency. They enable miniaturization and better bandwidth for wireless communication devices. Also, filters with metamaterials can be designed to allow only specific frequencies to pass while reflecting other frequencies; hence, cleaning the signal and minimizing interference.

Metamaterials are capable of overcoming the diffraction limit posed by conventional lenses, giving rise to superlenses that could provide

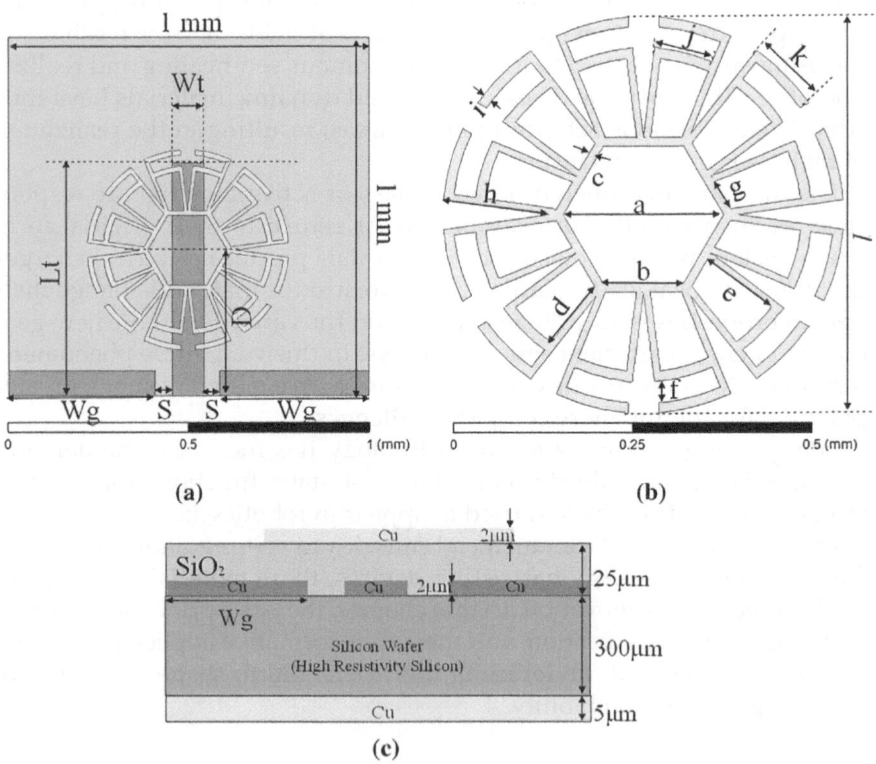

FIGURE 2.1
Top view of proposed antenna: (**a**) flower-metamaterial antenna and feed line (CPW), (**b**) flower-metamaterial design, and (**c**) cross-sectional view of proposed antenna [1].

high-resolution imaging. Applications in microscopy and medical imaging technologies include MRI and optical coherence tomography. An example of superlenses is illustrated in Figure 2.2.

Research on cloaking metamaterials aims at the creation of devices that would make objects undetectable for certain wavelengths of light. This can find applications in stealth technologies and advanced imaging systems (Figure 2.3).

Metamaterials can enhance the sensitivity of sensors to environmental changes, such as temperature, pressure, or chemical composition. They

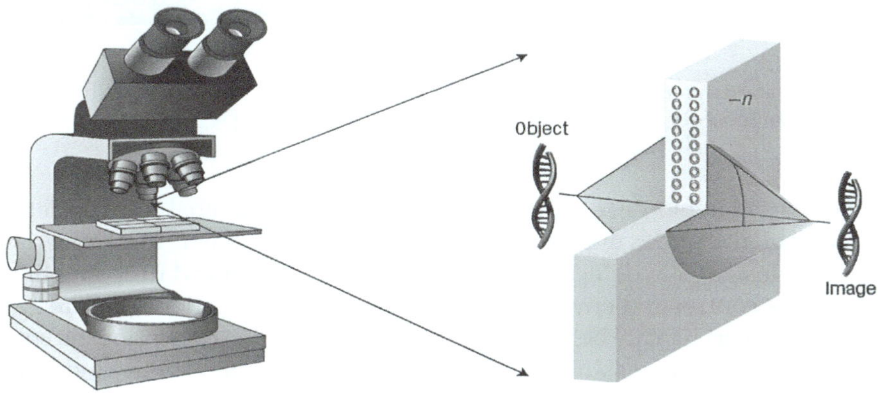

FIGURE 2.2
Metamaterials superlenses [2].

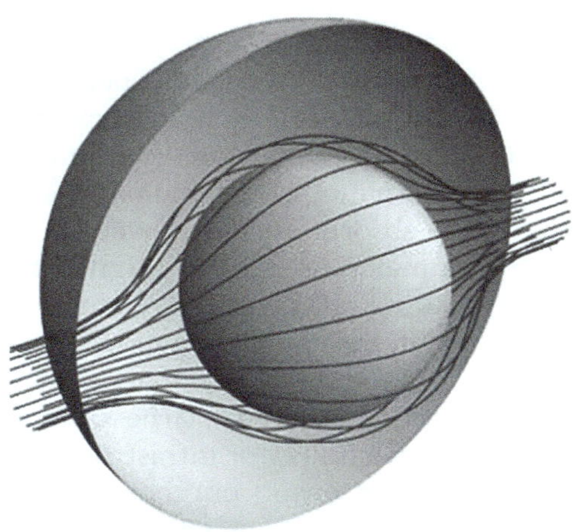

FIGURE 2.3
Metamaterial cloaking [2].

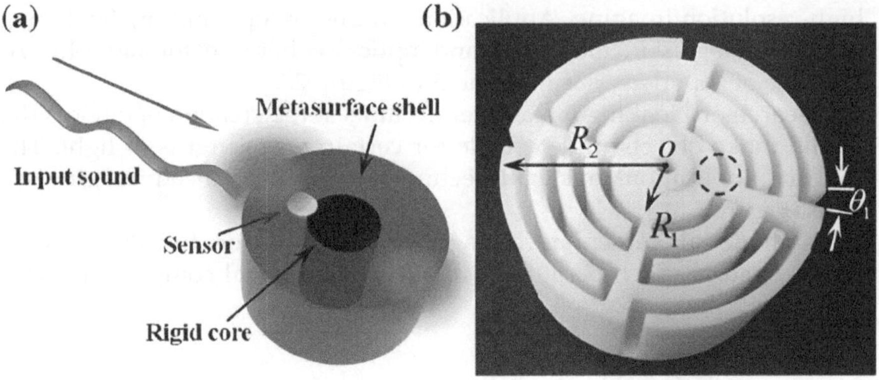

FIGURE 2.4
(a) Schematic diagram of metamaterial-enhanced acoustic directional sensing. (b) The fabricated sample of the core-shell-structured acoustic metaparticle [3].

can be used in biosensors for medical diagnostics, thus enabling the detection of biomolecules even at small concentrations. Acoustic metamaterials can be designed to detect sound waves, offering improved functionalities for noise control and environmental monitoring applications (Figure 2.4).

Metamaterials can be engineered to enhance the absorption of solar energy, hence increasing the efficiency of photovoltaic cells and solar thermal systems. A typical example is energy harvesting from the environment. Metamaterials are engineered to effectively convert the captured energy from the surrounding environment in forms like RF energy, sound, and mechanical vibration into electrical energy. For example, electromagnetic metamaterials are able to enhance the efficiency of antennas and systems of wireless energy transfer by focusing and amplifying RF signals. Acoustic metamaterials convert this energy to electrical energy through noise in the noisier the environment, harnessing the energy.

Another breakthrough involves solar energy technologies. The capability to manipulate light at the nanoscale by photonic metamaterials is being harnessed to increase the absorption and conversion efficiencies of photovoltaic cells. Such metamaterials can be designed to absorb a broader spectrum of sunlight, making solar panels even more effective in suboptimal lighting conditions. In the application of mechanical energy, metamaterials are embodied into systems to enhance the capture of kinetic energy. Such systems have the ability to harness vibrations or mechanical forces emanating from machinery or infrastructure and convert these into electrical energy to power sensors or other low-power devices. Their lightweight and customizable nature is particularly apt when metamaterials are integrated into wearable and portable devices. Metamaterials are also enabling improvements in wireless power transfer through the

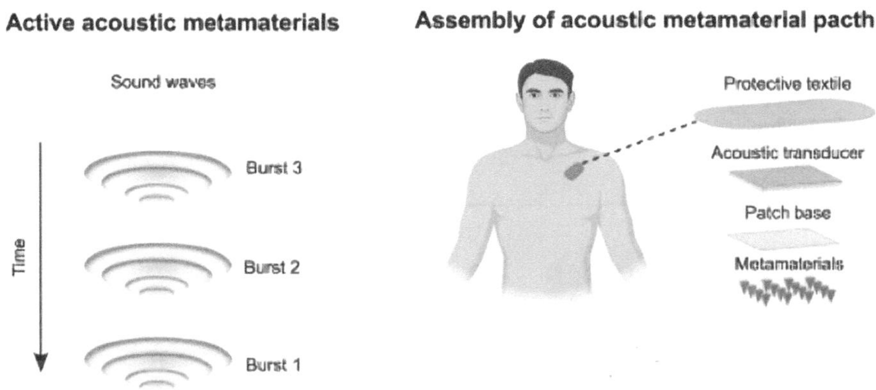

FIGURE 2.5
Schematics of the rapid and programmable transdermal drug delivery for timely controlling pharmacokinetics and managing the acute disease by an active acoustic metamaterial patch [4].

creation of resonant structures that tightly focus energy, reducing power loss and enhancing the efficiency of wireless charging systems for both electronic devices and electric vehicles.

Also, metamaterials can be used for thermal cloaking to manage the flow of heat for effective thermal management and enhancing energy efficiency in electronic devices.

Metamaterials can be used for the design of systems sensitive to certain stimuli like light or magnetic field for precise drug delivery, hence improving therapeutic efficiency in cancer treatment (Figure 2.5).

Metamaterials can enhance contrast and resolution in imaging techniques, hence improving diagnosis and treatment planning in healthcare. Mechanical metamaterials possess distinctive characteristics, including negative compressibility and superior energy absorption, rendering them ideal for protective equipment and lightweight armor. Metamaterials may serve in structures that change their mechanical properties in response to external excitations like vibrations or loads, therefore improving performance in a lot of engineering applications (Figure 2.6).

Acoustic metamaterials can be designed to alter sound waves – solutions related to noise mitigation in urban contexts or sound management in concert halls or theaters. These metamaterials would improve acoustic imaging techniques and would thereby lead to higher quality in medical ultrasound and nondestructive testing.

Metamaterials can be used for the construction of optical devices that are able to steer and focus beams of light very accurately, with obvious applications in optical communications and advanced imaging systems. A metamaterial can be designed to filter or polarize light selectively to

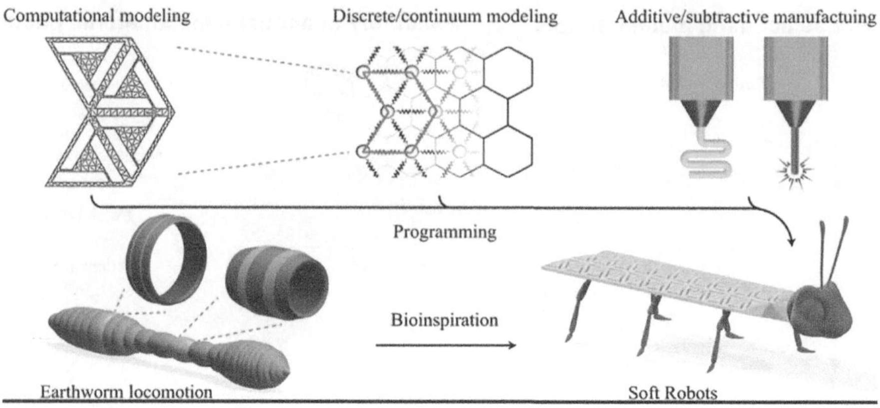

FIGURE 2.6
Bioinspired soft robots *via* adaptive mechanical metamaterials [5].

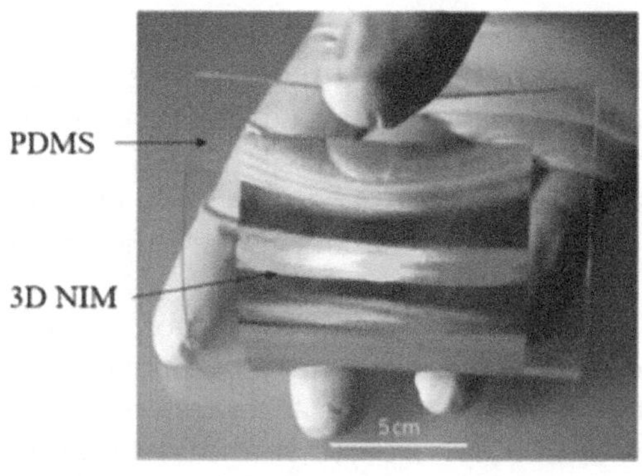

FIGURE 2.7
Polymer-based optical metamaterials. A large area (8.7 cm × 8.7 cm) flexible PDMS-based Ag negative index metamaterial (NIM) that shows negative index of refraction at optical frequencies. PDMS has the advantage of enabling direct pattern transfer, benefiting from PDMS's low surface energy [6].

extend the functions of optical devices like cameras, displays, and sensors (Figure 2.7).

The integration of metamaterials into soft robotic systems allows for adaptive and responsive motion, thus enhancing their ability to interact with complex environments. Through the use of the unique properties of active metamaterials, robotic systems can achieve complex shape changes, thus making them more adaptable in a variety of applications. Metamaterials represent one possible class of active material whose applications span telecommunications, imaging, sensing, energy harvesting, medicine, and beyond. Their ability to manipulate waves and respond adaptively to environmental stimuli empowers new solutions with performance and functionality enhancements in countless domains. As both metamaterials research and technology continue to progress, we can foresee applications becoming ever more diverse and significant in nature.

2.3 Application of Shape Memory Alloys as One Class of Active Material

Shape memory alloys are unique materials that return to a predetermined shape by simply exposing them to a particular temperature change or stress. Their ability to undergo significant deformation while returning to their original shape makes them useful in many applications. The following are the major uses of shape memory alloy as one class of active material:

Shape memory alloys find extensive applications in vascular stents due to their ability to expand at physiological temperatures. This enables minimal invasive implantation followed by expansion upon placement, thereby enhancing blood circulation. SMAs have the potential for applications in fixation devices and screws whose dimensions vary according to body temperature for enhanced comfort and bio-compatibility with the biological tissues (Figure 2.8).

Also, shape memory materials can be used in surgical instruments that change shape or function at a set temperature, thereby facilitating complex procedures. Shape memory alloys find their application in actuators for space applications, such as deployment of wing flaps and stabilization control in airplanes. In addition, due to their lightweight and compact form, they contribute to fuel efficiency.

In the automotive industry, SMAs can be integrated into systems that change the shape of vehicle components to optimize aerodynamics or comfort based on driving conditions.

Soft robotics rely on shape memory alloys to enable complex movements within a robotic system due to changing its form based on variations in temperature, mimicking organic movement. Shape memory alloys can be applied in robot grippers that adapt to the size and shape of various objects, therefore enhancing flexibility in automation.

Shape memory alloys can be integrated with fabrics to create clothing that reacts to temperature changes to enhance comfort and functionality related to outdoor or sportswear. SMAs can be used in wearable technology that adapts to the motion of the body, therefore enhancing

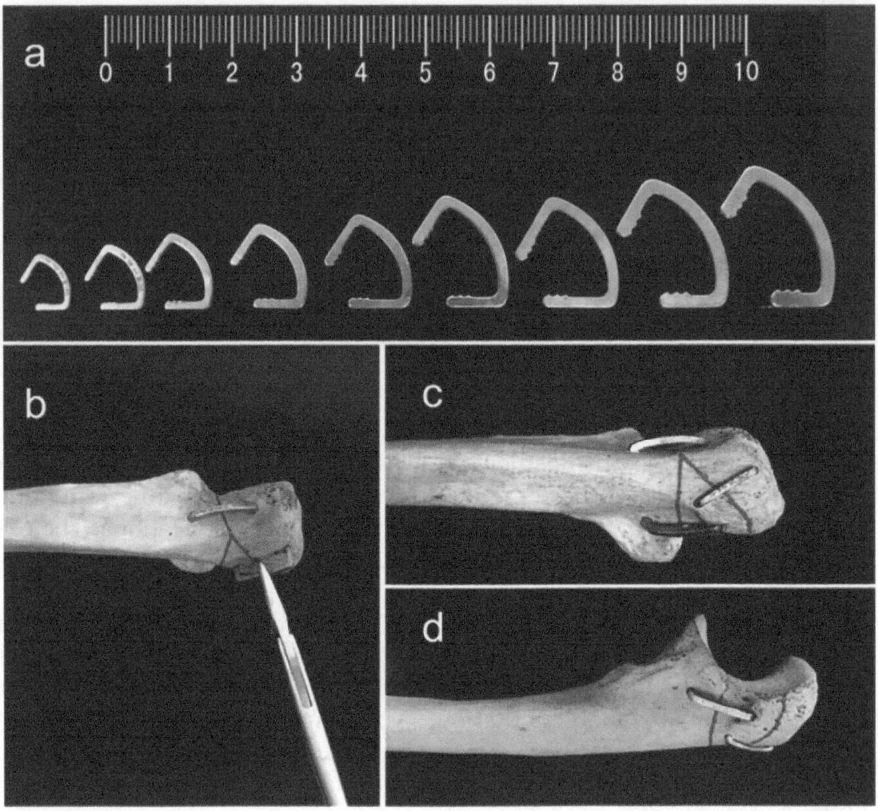

FIGURE 2.8
Different size of arched shape-memory connectors, which can be applied on different types of fractures, choose the most suitable arched shape-memory connector (a). Drill two holes on both sides of the fracture line and place the arched shape-memory connector vertically according to the fracture line (b). Finally, embed the compression arms and reheat in warm water to stimulate the reversion of the arms and waist back to their original shapes; several arched shape-memory connectors are used for the stable fixation of fractures and diagrams of the arched shape-memory connector fixation on specimen in AP (c) and lateral (d) views [7].

user experience in fitness trackers and health monitoring devices (Figure 2.9).

Shape memory alloys, SMAs, can be embedded within morphing wing structures that change shape in flight to optimize performance and fuel efficiency.

Shape memory alloys allow satellite antennas and/or reflectors to self-unfold and self-reconfigure in space.

Shape memory alloys are a multifaceted category of active materials utilized in many applications spanning biomedical devices, aircraft,

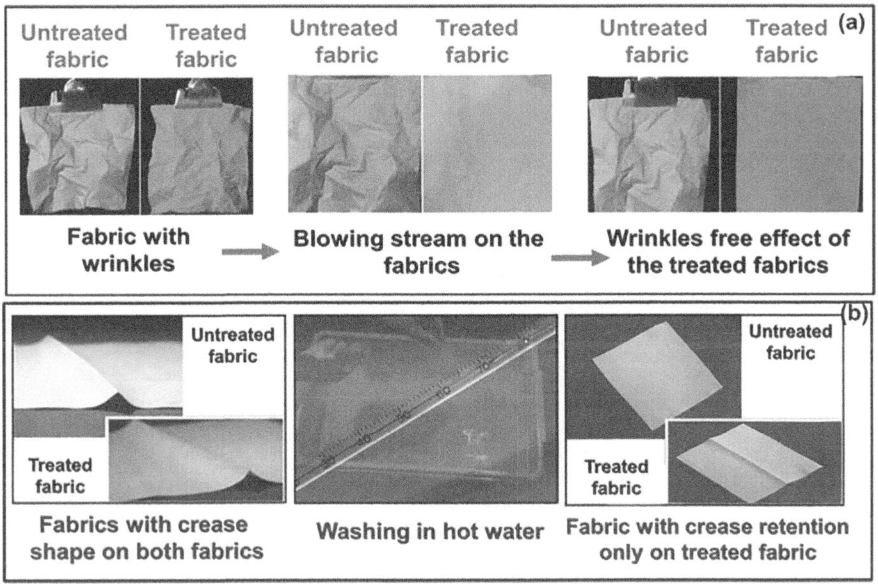

FIGURE 2.9
(a) Wrinkle-free effect and (b) crease and pattern retention ability of SMPU-treated cotton fabrics [8].

robotics, consumer electronics, civil engineering, textiles, and industrial systems. Their distinctive capacity to alter form in reaction to temperature or stress offers novel solutions that improve functionality, efficiency, and adaptability across several technologies. As research advances, the possible applications of shape memory alloys (SMAs) continue to proliferate, resulting in additional improvement in both established and novel technologies.

2.4 Application of Electroactive Polymers: A Class of Active Material

Electroactive polymers are a class of active materials that change shape or size when an electric field is applied. Their flexibility, light weight, and ability to achieve large deformation make them suitable for a wide range of applications. Some of the main applications of EAPs include the following: EAPs have been used in soft robotics for the development of flexible and adaptable actuators that are able to emulate natural movements.

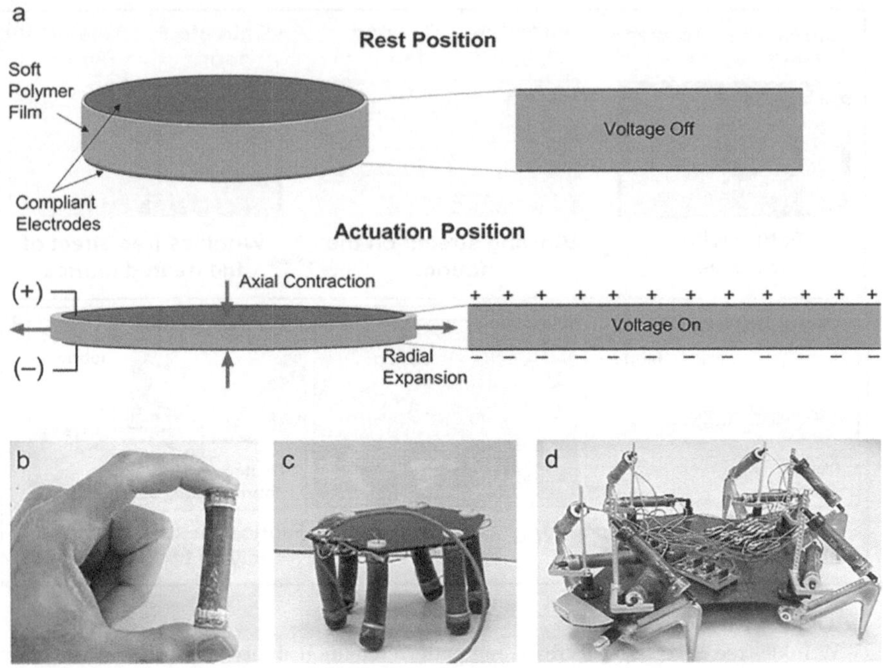

FIGURE 2.10
Electro-Active Polymers (EAPs). (a) Operating principle of dielectric elastomer actuators (DEAs). When a bias voltage is applied across an elastomer (soft polymer) film coated on both sides with compliant electrodes, Coulombic forces compress the film in the axial direction and expand it radially. (b) One degree of freedom multifunctional electroelastomer roll (MER) actuator fabricated with two prestrained films rolled around a compressed central spring (the spring is located at the center of the roll and covered by the films). (c) MERbot, a robot using a 2-DOF MER for each of its six legs. (d) FLEX 2 robot using MER actuators [9].

Their ability to deform due to electric fields allows for smooth and precise manipulation in robotic arms and grippers. EAPs act as artificial muscles in many applications, providing motion and force generation similar to that of real muscle tissues. This happens to be particularly helpful in prosthetics and exoskeletons (Figure 2.10).

EAPs can be integrated into touch-sensitive surfaces and devices to detect pressure and touches through changes in capacitance or resistance. EAPs can also be specifically designed to respond to environmental changes, like humidity or temperature, which makes them useful in weather stations and industrial processes (Figure 2.11).

EAPs can be applied to surgical devices that require fine control and dexterity, such as less-invasive devices that conform to the surgical site. Besides, electroactive polymers can be applied to implantable devices, which respond to electrical stimuli. They enable controlled drug delivery

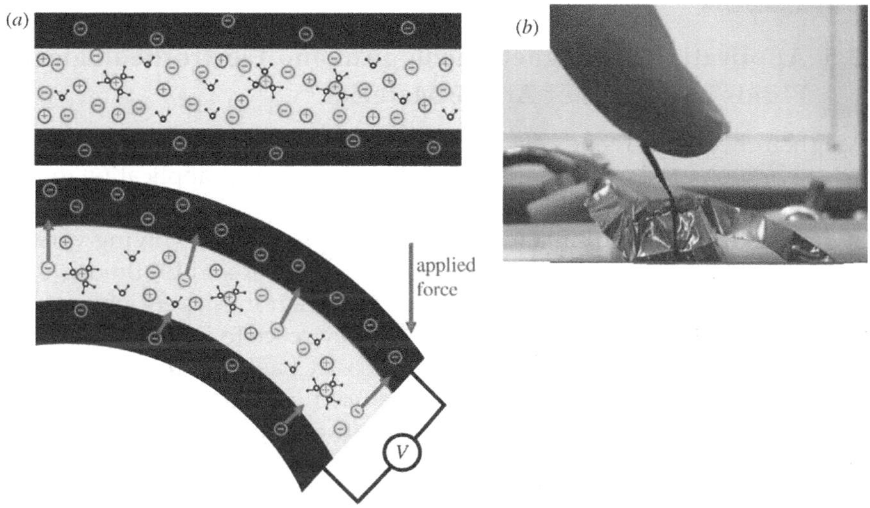

FIGURE 2.11
Schematic of (*a*) a bending trilayer sensor. In (*b*), a trilayer sensor is deflected by a finger [10].

or tissue stimulation. EAPs can convert mechanical energy generated from vibrations or motions into electrical energy; hence, they are suitable for small devices and sensors operating at remote locations. EAP-based devices can be designed to harvest energy from the environment – a common example is that supplied by body movement – to power sensors, thereby reducing dependency on external power sources.

EAP can be integrated into smart fabrics to create garments that change their properties in response to electrical stimulation, thus enabling adaptive fit or thermal regulation. EAP-enabled textiles can enable health monitoring applications, such as wearable sensors for the detection of physiological parameters and responding to human activity. EAPs can be applied in aerospace structures with the development of adaptive wings that change their shape during flight for better aerodynamics and fuel economy. EAPs are being used in spacecraft to deploy antennas or solar panels, allowing structures to unroll and change shape in response to electrical activation. EAPs are a multidisciplinary class of active material that can be applied in robotics, biomedical devices, energy harvesting, smart textiles, automotive systems, consumer electronics, and aerospace. The ability of changing shape or properties due to electrical excitation thus allows new approaches that enhance functionality and versatility in many technologies. Continued research in EAPs is expected to expand their range of applications, driving innovation in active materials and applications.

2.5 Application of Magnetorheological and Electrorheological Fluids as a Class of Active Material

Magnetorheological and electrorheological fluids are special liquids whose rheological characteristics change upon the application of an external magnetic or electric field, correspondingly. This unique reactivity makes them helpful in a wide range of applications involving various fields. The following are some important uses of MR and ER fluids:

Magnetorheological fluids find their application in suspension systems of vehicles to provide variable damping. They serve to increase ride comfort and stability by varying the fluid viscosity based on road conditions and driving dynamics. MR fluids can be used in clutches and brakes for variable torque transfer to facilitate smooth operation and performance variation with changing load conditions.

Magnetorheological fluids in structural dampers protect the edifices and bridges from seismic oscillations. During seismic events, these dampers can absorb and release energy, which indeed enhances structural resiliency. Electrorheological fluids can be used within the foundation systems to enable active control of soil-structure interaction and hence enhance stability in dynamic loading conditions. Magnetorheological and electrorheological fluids are used in soft robotic systems that require flexible and adaptive motions. These fluids allow for precise control over motion and force applied by flexible actuators. The use of MR or ER fluids allows actuators to achieve variable stiffness; thus, robots can change their rigidity to perform a variety of tasks in different environments.

MR fluids can be used in rehabilitation equipment, whose resistance is changed dynamically for personalized therapy for patients recovering from injury. ER fluids can be integrated into surgical instruments to provide variable adjustment of stiffness versus flexibility, which will allow precision with accuracy in minimally invasive procedures. Magnetorheological fluids are utilized in intelligent damping systems for furniture and appliances, offering customizable resistance to motion and enhancing user comfort (Figure 2.12).

Electrorheological fluids can be utilized in devices that deliver haptic feedback, facilitating tactile sensations in virtual reality systems and gaming controllers. Magnetorheological fluids can also be used in adaptive wing structures that change geometry or stiffness in response to aerodynamic conditions, thereby improving aircraft performance and fuel economy. MR dampers are applied in aerospace structures for the purpose of controlling unwanted vibrations and damping, hence increasing the stability of flying aircraft and spacecraft. MR and ER fluids are applied to various industrial machinery for active control of vibration and noise, which is expected to provide operational efficiency and prolong the

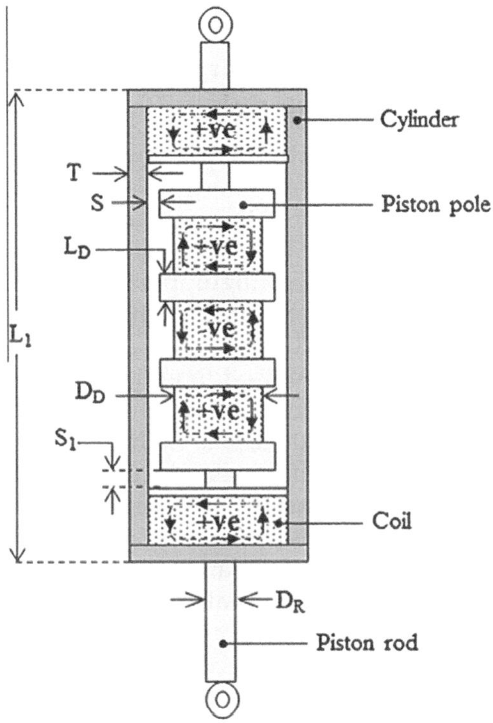

FIGURE 2.12
A sketch of the conceptual design of MR damper with the flux path illustration for the flow of flux around three coils, whose winding orientations alternate for each coil.

lifespan of equipment. In molding and casting operations, such fluids can be utilized where variable viscosity can allow better control of material flow and shaping.

Magnetorheological fluids can be applied within wind turbine control systems for the optimization of blade pitch and enhancement in energy capture in fluctuating wind. ER/MR fluids can be integrated into energy harvesting systems that convert mechanical energy from vibrations into electrical energy, which would make sensors and devices self-sustaining.

MR/ER fluids are a class of active multifunctional material applied in such fields as automotive, civil engineering, robotics, medical devices, consumer products, aerospace, industrial machinery, and energy systems. Changes in rheological properties induced by external fields create opportunities for original solutions with enhanced performance and adaptability, as well as increased efficiency in a wide range of technologies. The advances of research will further expand the range of applications using MR/ER fluids, which in turn drive innovation in active materials and their applications.

2.6 Application of Shape Memory Polymer as a Type of Active Material

Shape memory polymers (SMPs) are polymers capable of returning to their original shape after being deformed. This "memory" originates from the mechanical energy stored throughout the material's cooling process. The shape memory effect denotes a material's ability to revert from a deformed state to its original configuration upon the application of an external stimulus. Shape memory effect (SME) was identified in 1932 by Chang and Read in alloys. This behavior was subsequently found in polymers. The SME principle can be used to alloys, polymers, ceramics, and gels. In many SMPs, the mechanism turns out to be biphasic, which consists of a transient phase and a stable phase. The temporary phase becomes malleable with the application of heat or any other stimulus; subsequently, this allows the material to revert to its permanent configuration. SMPs have been designed to exhibit dynamic responses to environmental stimuli, including temperature, light, pH, and solvents.

Shape memory polymers use thermal energy to activate shape memory. Such a system enables the polymer matrix to change from the glassy state to the rubbery state, which then can recover its original configuration. The temperature-based activation is quite efficient; however, it either requires direct heating or changes in the ambient, which can create limitations in certain applications. SMPs are composed of one polymer matrix, which can be chosen based on the desired mechanical and thermal properties. Both thermoplastic and thermoset polymers are in widespread use; both types of polymers exhibit very predictable behavior in the absence of additional fillers. It is also a fact that shape memory polymers are limited in terms of the speed at which they can actuate; their recovery process is bound by the rate of thermal input applied, and using environmental heat sources could yield a relatively slow actuation process. SMPs exhibit a poor self-healing capability; most of the time, supplementary processes for the healing of damage are necessary, which often involve direct heating. In SMPs, there is considerable freedom in material design because there are fewer constraints on the composition of the polymer matrix. Thermoplastic and thermoset polymers are available with various glass transition or melting temperatures for applications in a wide temperature range. SMPs are more cost-effective and easy to fabricate on a large scale, as they do not require the usage of nanoparticles, and also their industries are already well established. Shape memory polymers have a wide range of applications whenever the principle of thermal activation can be applied. Examples of biomedical applications, such as stents, implants, automotive parts like seals and sensors, and aerospace deployable structures, are some of the important uses. Currently, research is being carried out on their application to smart textiles and different adaptive materials. The following are some of the major uses of SMPs:

SMPs are used in vascular stents and in orthopedic implants, which could expand or change shape at body temperature to facilitate their minimally invasive placement and to enhance tissue integration (Figure 2.13).

SMPs can be designed to release drugs in a controlled manner upon receiving stimuli from changes in temperature, pH, or other environmental variables, thus resulting in an enhancement of therapeutic efficacy (Figure 2.14).

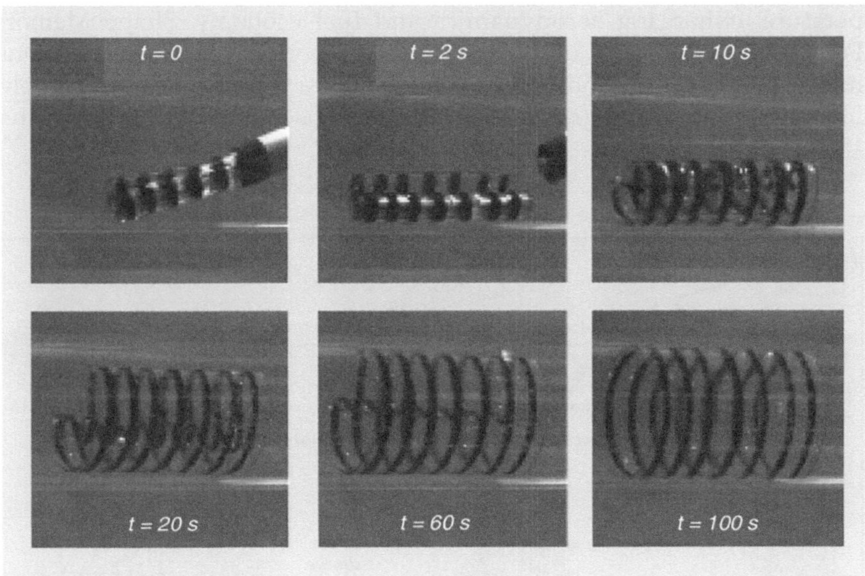

FIGURE 2.13
Insertion of a transparent shape-memory polymer stent through a 22-mm catheter [11].

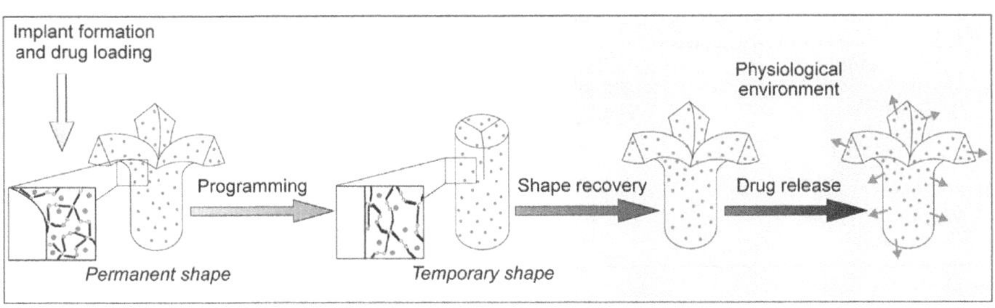

FIGURE 2.14
Concept of programming, shape recovery, and drug release of drug-loaded SMP devices for biomedical applications [12].

SMPs can be included in wound dressings that change properties in response to body temperature – for example, shape or permeability – to provide improvements in healing and comfort. SMPs are embedded in fabrics that can change shape or property in response to temperature for adaptive fitting or heat management in clothing (Figure 2.15).

SMPs can provide adaptive support or cushioning for sport equipment to respond to the specific movements and body temperature of athletes. Also, SMPs can be integrated into the seats of cars that adjust their shape for enhanced comfort based on the occupant's posture and temperature. SMPs can be used in car grilles that move with respect to engine temperature, enhancing aerodynamics and fuel economy. Shape Memory Polymers enable soft robotic systems to perform complex motions, mimic biological motions, and expand into various applications involving manipulation and grasping. SMPs can be used in robotic systems that

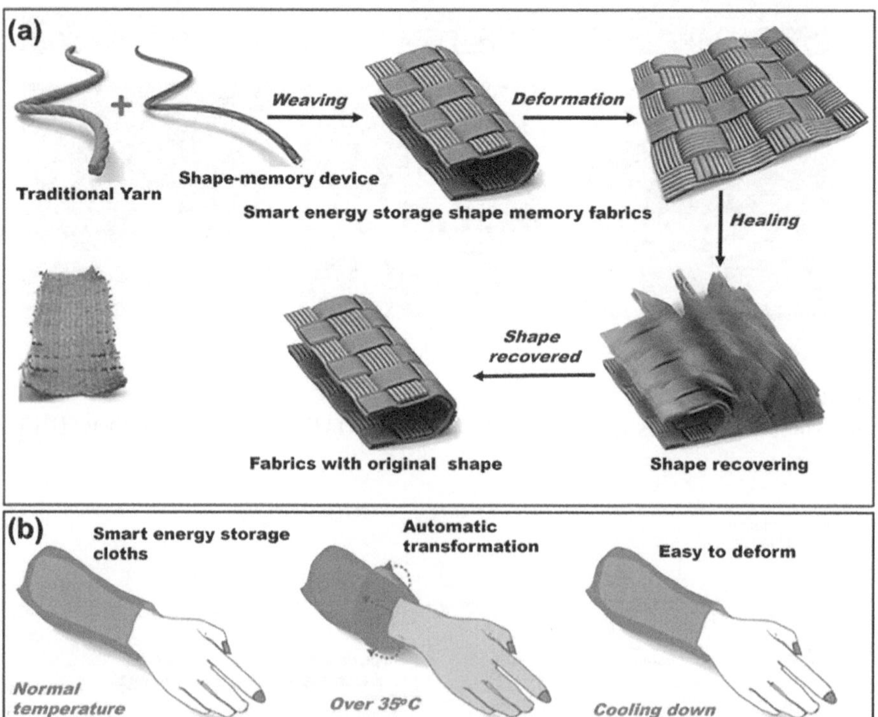

FIGURE 2.15
(a) Schematic demonstration of the shape memory SC woven with traditional fabric and the fabrication of a shape-recoverable smart textile and (b) schematic demonstration of this smart shape-memory textile used in a smart cloth [13].

require transformable architecture and thus can change their shape for different tasks or conditions.

Shape memory polymers can be embedded in consumer products to help them return to their original form after deformation, thereby enhancing durability and usability (Figure 2.16).

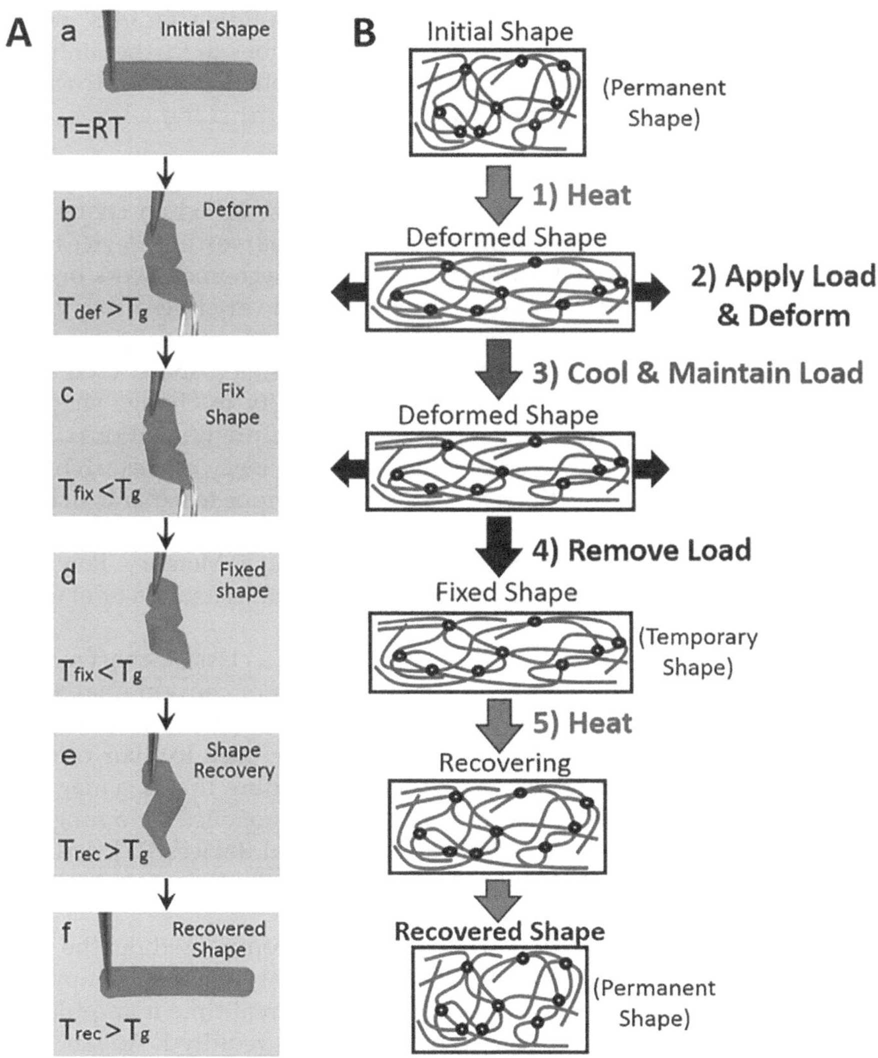

FIGURE 2.16
Conventional Tg-based shape memory cycle (SMC): initial shape at room temperature, (a) shape at T > Tg, (b) fixed shape at room temperature, (c–d) and shape recovery process at recovery temperature TR (e–f) [14].

SMPs can be applied in such a way that the packaging itself changes shape or configuration in response to temperature, providing better protection of the product and ease of use for the user. SMPs can be used in aerospace structures, such as morphing wings that change their shape during flight in order to optimize aerodynamics and save fuel. Also, in space, SMPs allow for the self-deployment of antennas or solar panels, for example, which come folded and deploy themselves in operation. Shape-shifting SMPs could be integrated into building façades, which change shape depending on temperature fluctuations to further enhance energy efficiency and aesthetics. SMPs can be applied in those structural elements subjected to dynamic load, enhancing further the seismic resistance of edifices.

Shape memory polymers are thermally actuated in energy applications with the ability to enable devices for performance based on changes in temperature. SMPs can be included in energy harvesting devices that transform environmental heat fluctuations into mechanical work or electrical energy. Shape memory polymers represent a very interesting group of materials that have found applications in biomedical devices, textiles, automotive systems, robotics, consumer products, aerospace, civil engineering, and energy solutions. Due to their ability to change shape in response to temperature or other stimuli, completely new solutions can be enabled that enhance utility, adaptability, and efficiency. As research continues, the range of applications for SMPs will continue to expand, driving further innovation of active materials and applications.

Also, as a subsection of SMPs, Magnetic Shape Memory Polymers (MSMPs) have gained lots of attention in recent years. Next is a brief introduction to MSMPs.

Magnetic shape memory polymers (MSMPs) are a class of smart materials that embody the thermo-reversible properties of conventional shape memory polymers, ameliorated with enhanced magnetic responsivity. These can undergo shape deformation and come back to their original configuration [15] upon exposure to external stimulus, that is, a magnetic field in this case. The development of MSMPs has opened up a range for new applications in drug delivery [16], biomedical devices [17], artificial muscles [18], and actuators [19] that require remote non-contact control [20, 21].

An option for remote actuation allows external control without the need for direct accessibility, which makes MSMP highly relevant to applications in medicine, robotics, or other areas where real-time manipulation with use of magnetic field-sensing information is required. Advances of MSMPs with respect to classical SMP [22] and other smart materials are non-contact actuation: using magnetic fields [23] to enable controlling functions without actively making contact, thus reducing the reliance on wires and physical response. It is very helpful in environments that are either unreachable or difficult; precision: magnetic fields could be applied very precisely; hence, changes in shape can be thoroughly controlled. To activate parts of the material in a highly specific manner, the magnetic field could be changed in intensity, frequency, or orientation; biocompatibility, which is the use of biocompatible magnetic nanoparticles, such as iron oxide, renders MSMPs suitable for medical applications, especially in implants and devices that engage with human tissue; fast response, where MSMPs can immediately revert to their original shape when exposed to a magnetic field. Thus, they are fit for applications that require speed and instantaneous actuation.

On the other hand, such potentials face considerable obstacles in the advancement and application of magnetic shape memory polymers. For example, a homogeneous distribution [24] of magnetic nanoparticles in the polymer matrix poses a formidable technological challenge, where poor distributions result in an inconsistent actuation response and poor material performance. Long-term reliability demands optimization of the thermal stability and effectiveness of the magnetic particles within successive cyclic actuation. Cost-effective methods of manufacturing and scalability remain key issues, especially in industrial needs that call for mass production. In addition, potential biocompatibility and safety issues related to MSMPs, particularly in medical use, should be carefully considered in order to safely introduce these materials into healthcare technology.

The major principle behind the functioning of MSMPs relies on the heating effect developed by magnetic nano particles (MNPs) within an alternating magnetic field (AMF) [25]. This technique is called magnetic induction heating and, locally or often globally, increases the temperature of the polymer matrix, which then acts on a phase transition and returns the polymer to its original configuration [26]. Complementing the thermal effects, magneto-mechanical phenomena arise when a static magnetic field applies a mechanical force to magnetic particles for the purpose of causing deformation or even direct actuation. The underlying magneto-thermal principle depends on two key factors: nanoparticle heating, which refers to the process whereby magnetic nanoparticles absorb energy from an alternating magnetic field and subsequently release it as thermal energy, and shape recovery, which occurs where the heat applied starts a transformation of the polymer from its temporary configuration to permanent form.

The fact that morphology is controlled by the magnetic field in real time provides enormous advantages in applications requiring non-contact actuation over traditional shape memory polymers, whose triggering relies exclusively either on ambient heating or on direct contact [27]. A schematic description of MSMPs has been displayed in Figure 2.17.

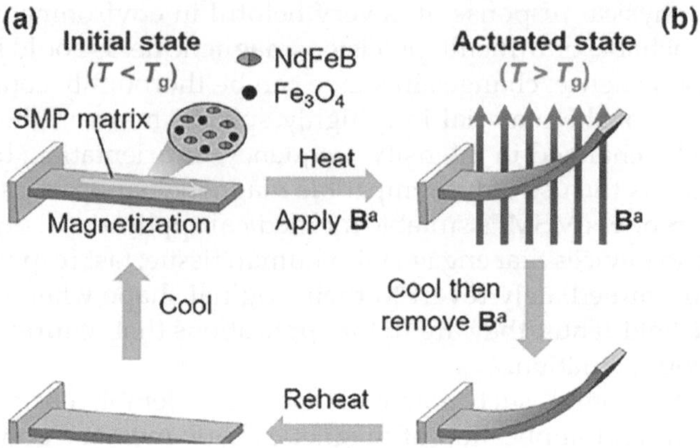

FIGURE 2.17
The operating principles and thermomechanical properties of magnetic-shape memory polymers (M-SMPs). (a) The M-SMP is created by integrating Fe_3O_4 and NdFeB particles into the SMP matrix. (b) Storage moduli and tanδ (the ratio of loss modulus to storage modulus) as a function of temperature [28].

References

1. Olan-Nuñez, K.N., and R.S. Murphy-Arteaga, A novel metamaterial-based antenna for on-chip applications for the 72.5–81 GHz frequency range. *Scientific Reports*, 2022. 12(1): p. 1699.
2. Aydın, K., *Characterization and Applications of Negative-Index Metamaterials*. 2008: Bilkent Universitesi (Turkey).
3. Zhu, X., et al., Deep-subwavelength-scale directional sensing based on highly localized dipolar mie resonances. *Physical Review Applied*, 2016. 5(5): p. 054015.
4. Xu, J., et al., Acoustic metamaterials-driven transdermal drug delivery for rapid and on-demand management of acute disease. *Nature Communications*, 2023. 14(1): p. 869.
5. Khajehtourian, R., and D.M. Kochmann, Soft adaptive mechanical metamaterials. *Frontiers in Robotics and AI*, 2021. 8: p. 673478.
6. Chen, Y., B. Ai, and Z.J. Wong, Soft optical metamaterials. *Nano Convergence*, 2020. 7(1): p. 18.
7. Xia, D., et al., Application of a novel shape-memory alloy concentrator in displaced olecranon fractures: a report of the technique and mid-term clinical results. *Journal of Orthopaedic Surgery and Research*, 2020. 15: p. 1–8.
8. Hu, J., et al., A review of stimuli-responsive polymers for smart textile applications. *Smart Materials and Structures*, 2012. 21(5): p. 053001.

9. Higueras-Ruiz, D.R., et al., What is an artificial muscle? A comparison of soft actuators to biological muscles. *Bioinspiration & Biomimetics*, 2021. 17(1): p. 011001.
10. Wang, T., et al., Electroactive polymers for sensing. *Interface Focus*, 2016. 6(4): p. 20160026.
11. Lendlein, A., et al., Shape-memory polymers as a technology platform for biomedical applications. *Expert review of Medical Devices*, 2010. 7(3): p. 357–379.
12. Wischke, C., et al., Evaluation of a degradable shape-memory polymer network as matrix for controlled drug release. *Journal of Controlled Release*, 2009. 138(3): p. 243–250.
13. Thakur, S., Shape memory polymers for smart textile applications, in *Textiles for Advanced Applications*. 2017: InTech. p. 10.
14. Hornat, C.C., and M.W. Urban, Shape memory effects in self-healing polymers. *Progress in Polymer Science*, 2020. 102: p. 101208.
15. Lendlein, A., and S. Kelch, Shape-memory polymers. *Angewandte Chemie International Edition*, 2002. 41(12): p. 2034–2057.
16. Vakil, A.U., M. Ramezani, and M.B.B. Monroe, Magnetically actuated shape memory polymers for on-demand drug delivery. *Materials*, 2022. 15(20): p. 7279.
17. Yarali, E., et al., Magneto-/electro-responsive polymers toward manufacturing, characterization, and biomedical/soft robotic applications. *Applied Materials Today*, 2022. 26: p. 101306.
18. Bastola, A.K., and M. Hossain, The shape – morphing performance of magnetoactive soft materials. *Materials & Design*, 2021. 211: p. 110172.
19. Kim, J., et al., Programming magnetic anisotropy in polymeric microactuators. *Nature Materials*, 2011. 10(10): p. 747–752.
20. Kumar, U.N., et al., Non-contact actuation of triple-shape effect in multiphase polymer network nanocomposites in alternating magnetic field. *Journal of Materials Chemistry*, 2010. 20(17): p. 3404–3415.
21. Vialle, G., et al., Remote activation of nanomagnetite reinforced shape memory polymer foam. *Smart Materials and Structures*, 2009. 18(11): p. 115014.
22. Liu, T., et al., Stimulus methods of multi-functional shape memory polymer nanocomposites: a review. *Composites Part A: Applied Science and Manufacturing*, 2017. 100: p. 20–30.
23. Sánchez, C.P., et al., Review of thermoresponsive electroactive and magnetoactive shape memory polymer nanocomposites. *ACS Omega*, 2022. 7(45): p. 40701–40723.
24. Phetnoi, N., et al., Fast magnetic-responsive shape memory composites from bio-based benzoxazine/polyglycerol polyglycidyl ether copolymers highly filled with iron oxide nanoparticles. *Composites Part A: Applied Science and Manufacturing*, 2024. 186: p. 108398.
25. Buckley, P.R., et al., Inductively heated shape memory polymer for the magnetic actuation of medical devices. *IEEE Transactions on Biomedical Engineering*, 2006. 53(10): p. 2075–2083.

26. Galarreta-Rodriguez, I., et al., Magnetically activated 3D printable polylactic acid/polycaprolactone/magnetite composites for magnetic induction heating generation. *Advanced Composites and Hybrid Materials*, 2023. 6(3): p. 102.
27. Zhao, F., et al., Photothermal-responsive shape-memory magnetic helical microrobots with programmable addressable shape changes. *ACS Applied Materials & Interfaces*, 2023. 15(21): p. 25942–25951.
28. Lu, L., S. Wu, and R.R. Zhao, Mechanics of magnetic-shape memory polymers. *Journal of the Mechanics and Physics of Solids*, 2024: p. 105742.

3

Vibration of Active Material Beams

3.1 Background

The primary aim of this section is to analyze the vibrational properties of a nanobeam displaying auxetic behavior, facilitated by the inclusion of a magnetostrictive substance known as Terfenol-D. The nanobeam is hypothesized to consist of three separate layers, with the middle layer made of a magnetostrictive material and the outer layers composed of auxetic substances. The higher-order parabolic shear deformation beam theory is employed to derive the kinematic relations. The findings of this study may significantly enhance the comprehension and development of nano-systems, specifically nano-sensors and nano-actuators. Moreover, the findings from this study may provide a foundational framework for subsequent research. The exceptional properties exhibited by metamaterials can be ascribed to their distinctive structure, realized by the amalgamation of diverse materials arranged in a repeated micro-unit configuration. The Poisson's ratio for most materials is typically positive; however, a particular production procedure for metamaterials has produced a negative Poisson's ratio. Auxetics, originating from the Greek word "auxetikos," signifying "to grow," denote a category of metamaterials characterized by a negative or zero Poisson's ratio [1]. Some mechanical materials exhibit superior characteristics, including compression and shear resistance, energy absorption, sound insulation, synclastic behavior, anisotropy, high elasticity in auxetic materials adhering to Hooke's law, and significant damping resistance, attributable to their negative Poisson's ratio (NPR) [1–6]. Currently, significant attention has been directed toward magnetostrictive materials, including Terfenol-D and Galfenol, owing to their distinctive properties. The unique characteristic of these intelligent materials is in their ability to transform energy from one form, such as magnetic, electric, or mechanical, into other energy forms. Materials like Terfenol-D possess exceptional magnetostrictive properties, allowing them to generate significant magnetic strains. As a result, these materials

DOI: 10.1201/9781003540731-3

are very appropriate for application in highly sensitive sensors. The examination of the vibrational characteristics of magnetostrictive materials is critically significant due to their inherent qualities.

3.2 Theoretical Framework and Formulation

This study examines the structural configuration of a nanobeam situated on an elastic substrate, specifically known as the Winkler-Pasternak model. The nanobeam is defined by its dimensions in the x, y, and z axes, represented as length L, width b, and total thickness H, respectively. The Cartesian coordinate system, illustrated in Figure 3.1, has been employed in the present investigation.

3.2.1 Displacement Field

The displacement field can be articulated as follows, based on the principles of Reddy's higher-order shear deformation beam theory [7]:

$$U_x(x,z,t) = u_0(x,t) + z\phi(x) - \alpha z^3\left(\phi + \frac{\partial w_0}{\partial x}\right) \tag{3-1}$$

$$V_y(x,z,t) = 0 \tag{3-2}$$

$$W_z(x,z,t) = w_0(x,t) \tag{3-3}$$

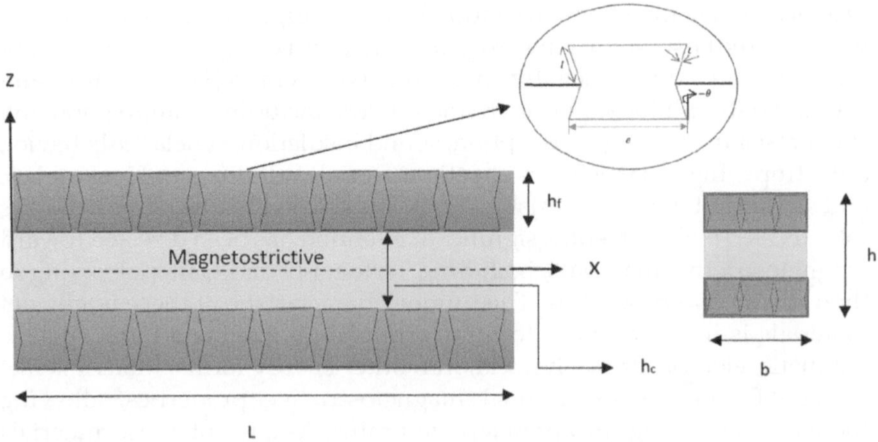

FIGURE 3.1
Configuration of the sandwich composite.

The constants u_0 and w_0 denote the longitudinal and transverse displacements, respectively. Furthermore, ϕ represents the rotation of the cross-section at each point along the neutral axis.

3.2.2 Auxetic Materials

The possible ramifications concerning the efficient mechanical properties of the auxetic core are as follows [2]:

$$E_{11}^{a} = E_{Al}\left(\frac{t}{l}\right)^{3}\frac{\cos\theta}{\left(\frac{e}{l}+\sin\theta\right)\sin^{2}\theta} \tag{3-4}$$

$$E_{22}^{a} = E_{Al}\left(\frac{t}{l}\right)^{3}\frac{\left(\frac{e}{l}+\sin\theta\right)}{\cos^{3}\theta} \tag{3-5}$$

$$G_{12}^{a} = E_{Al}\left(\frac{t}{l}\right)^{3}\frac{\left(\frac{e}{l}+\sin\theta\right)}{\left(\frac{e}{l}\right)^{2}\left(1+\frac{2e}{l}\right)\cos\theta} \tag{3-6}$$

$$v_{12}^{a} = \frac{\cos^{2}\theta}{\left(\frac{e}{l}+\sin\theta\right)\sin\theta} \tag{3-7}$$

$$\rho_{a} = \rho_{Al}\frac{\left(\frac{t}{l}\right)\left(\frac{e}{l}+2\right)}{2\cos\theta\left(\frac{e}{l}+\sin\theta\right)} \tag{3-8}$$

The equations outlined in Eq. (3–4) to Eq. (3–8) facilitate the assessment of the mechanical properties of the utilized auxetic core. The rib thickness of the auxetic cell is denoted by the variable t, its horizontal rib length by e, and its vertical inclined rib length by l. Two effective auxetic core moduli exist: the Young's moduli E_{11}^{a} and E_{22}^{a}, along with the shear moduli G_{11}^{a} and G_{22}^{a}. Additionally, both v_{11}^{a} and v_{22}^{a} can be seen as effective auxetic modifications of Poisson's ratio.

3.2.3 Eringen's Nonlocal Theory

Eringen's nonlocal theory has been applied in situations where the significance of geometric dimensions is paramount. The theoretical framework indicates that the stress at a particular reference point (x) in an elastic continuum is affected by the strain at point x and the strains of all adjacent points. Eringen's nonlocal theory is commonly applied in micro and nanoscale contexts. The nonlocal stress tensor at position x can be expressed as [58]:

$$\left(1-\mu^2\nabla^2\right)\sigma^{nl}=\sigma^{cl} \tag{3-9}$$

where σ^{nl} and σ^{cl} denote the nonlocal stress tensor and the classical stress tensor, respectively. Furthermore, ∇^2 denotes the Laplacian operator. Furthermore, μ = e0, which represents the nonlocal parameter.

3.2.4 Stress-Strain Relationship for the Magnetostrictive Core Layer

Further work may elucidate the link between stress, strain, and magnetic properties within the core layer [7]:

$$\left(1-\mu^2\nabla^2\right)\begin{Bmatrix}\sigma_{xx}^{c,nl}\\ \tau_{xz}^{c,nl}\end{Bmatrix}=\begin{bmatrix}C_{11}^{c} & 0\\ 0 & C_{55}^{c}\end{bmatrix}\begin{Bmatrix}\varepsilon_{xx}^{c}\\ \gamma_{xz}^{c}\end{Bmatrix}-H_{zz}\begin{Bmatrix}\overline{e}_{31}\\ 0\end{Bmatrix} \tag{3-10}$$

Where $\sigma_{ij},\varepsilon_{ij}\left(i,j=x,y,z\right)$ denote stress and strain constants, respectively. Furthermore, H_{zz} denotes a magnetic field. Furthermore, it is important to acknowledge that C_{ij} denotes the elastic stiffness coefficient:

$$C_{11}^{c}=\frac{E_c}{1-\vartheta_c^{\,2}},\; C_{55}^{c}=\frac{E_c}{2\left(1+\vartheta_c\right)} \tag{3-11}$$

E_c and v_c represent the modulus of elasticity and Poisson's ratio of the core layer, respectively, whereas $\overline{e}_{ij}$ signifies the converted magnetostrictive moduli, which can be expressed as [7]:

$$\begin{aligned}\overline{e}_{31}&=e_{31}\cos^2\theta+e_{32}\sin^2\theta\\ \overline{e}_{32}&=e_{32}\cos^2\theta+e_{31}\sin^2\theta\\ \overline{e}_{34}&=(e_{31}-e_{32})\sin\theta\cos\theta\end{aligned} \tag{3-12}$$

3.2.5 Velocity Feedback Coefficient

The magnetic field $\left(H_{zz}\right)$ can be derived using the subsequent equation [7]:

$$\begin{aligned}H_{zz}&=k_c I\left(x,t\right)\\ I\left(x,t\right)&=C\left(t\right)\frac{\partial w_0\left(x,t\right)}{\partial t}\end{aligned} \tag{3-13}$$

Here, k_c denotes the coil constant, which is contingent upon the coil's characteristics, including its width, radius, and the number of turns. $I(x,t)$ and $C(t)$ represent coil current and coil gain, respectively. The control gain is regarded as constant. $K_c C(t)$ represents the velocity feedback gain. The relationship between coil constant and coil characteristics can be articulated as:

$$K_c = \frac{n_c}{\sqrt{b_c^2 + 4r_c^2}} \tag{3-14}$$

In the aforementioned equation, n_c, b_c, and r_c represent the number of turns of the coil, the width of the coil, and the radius of the coil, respectively. The material properties of the magnetostrictive core layer [9] and auxetic facesheets [4] are presented next.

The non-zero stresses of the core layer can be categorized as follows:

$$\varepsilon_{xx}^{c} = \varepsilon_{xx}^{c} + z\kappa_{xx}^{c} + z^3\chi_{xx}^{c} \quad \gamma_{xz}^{c} = \varepsilon_{xz}^{c} + z^2 X_{xz}^{c} \tag{3-15}$$

in which

$$\varepsilon_{xx}^{c} = \frac{\partial u_0}{\partial x}, \kappa_{xx}^{c} = \frac{\partial \phi}{\partial x}, \chi_{xx}^{c} = -\alpha\left(\frac{\partial \phi}{\partial x} + \frac{\partial^2 w_0}{\partial x^2}\right)$$

$$\varepsilon_{xz}^{c} = \frac{\partial w_0}{\partial x} + \phi, X_{xz}^{c} = -\beta\left(\frac{\partial w_0}{\partial x} + \phi\right) \tag{3-16}$$

3.2.6 Stress-Strain Relationship of the Auxetic Facesheets

The mathematical formulation of the relationship between stress and strain for auxetic material facesheets can be articulated as:

$$\left(1 - \mu^2\nabla^2\right)\begin{Bmatrix} \sigma_{xx}^{a,nl} \\ \tau_{xz}^{a,nl} \end{Bmatrix} = \begin{bmatrix} C_{11}^a & 0 \\ 0 & C_{55}^a \end{bmatrix}\begin{Bmatrix} \varepsilon_{xx}^a \\ \gamma_{xz}^a \end{Bmatrix} \tag{3-17}$$

The notion of elastic functions may be delineated within specific situations.

$$C_{11}^a = \frac{E(z)}{1 - \vartheta_a^2} \quad C_{55}^a = \frac{E(z)}{2(1 + \vartheta_a)} \tag{3-18}$$

in which E_a and v_a are defined as the elastic modulus and Poisson's ratio of the face sheet, respectively.

The strains of the upper and lower auxetic facesheets, which are nonzero, may be defined as follows:

$$\varepsilon_{xx}^{a} = \varepsilon_{xx}^{a} + z\kappa_{xx}^{a} + z^{3}\chi_{xx}^{a} \quad \gamma_{xz}^{a} = \varepsilon_{xz}^{a} + z^{2}X_{xz}^{a} \tag{3-19}$$

in which

$$\varepsilon_{xx}^{a} = \frac{\partial u_0}{\partial x}, \kappa_{xx}^{a} = \frac{\partial \phi}{\partial x}, \chi_{xx}^{a} = -\alpha\left(\frac{\partial \phi}{\partial x} + \frac{\partial^2 w_0}{\partial x^2}\right)$$

$$\varepsilon_{xz}^{a} = \frac{\partial w_0}{\partial x} + \phi, X_{xz}^{a} = -\beta\left(\frac{\partial w_0}{\partial x} + \phi\right) \tag{3-20}$$

where α and β are defined as: $\alpha = \frac{4}{3h^2}, \beta = \frac{4}{h^2}$

3.3 Governing Equation

The governing equations of equilibrium can be derived by integrating the displacement gradients by parts and subsequently setting the coefficients δu_0, δw_0, and $\delta \phi$ to zero separately. The initial articulation of Hamilton's concept can be expressed as follows [8]:

$$\int_0^t \delta(U - T + V)\, dt = 0 \tag{3-21}$$

In this context, U, T, V represent strain energy, kinetic energy, and external work, respectively.

3.3.1 Strain Energy

The initial variation expression for the strain energy of the nanobeam can be theoretically articulated as follows [7]:

$$\delta U^{c} = \int_{v}\left(N_{xx}^{c}\left(\delta\varepsilon_{xx}^{c}\right) + M_{xx}^{c}\left(\delta\kappa_{xx}^{c}\right) + G_{xx}^{c}\left(\delta\chi_{xx}^{c}\right) + N_{xz}^{c}\left(\delta\varepsilon_{xz}^{c}\right) + P_{xz}^{c}\left(\delta X_{xz}^{c}\right)\right) dV$$

$$\delta U^a = \int_v \left(N_{xx}^a \left(\delta\varepsilon_{xx}^a \right) + M_{xx}^a \left(\delta\kappa_{xx}^a \right) + G_{xx}^a \left(\delta\chi_{xx}^a \right) + N_{xz}^a \left(\delta\varepsilon_{xz}^a \right) + P_{xz}^a \left(\delta X_{xz}^a \right) \right) dV \tag{3-22}$$

in which,

$$\left(N_{ij}^{\ c}, M_{ij}^{\ c}, P_{ij}^{\ c}, G_{ij}^{\ c} \right) = \int_{-\frac{h_c}{2}}^{\frac{h_c}{2}} \sigma_{ij}^{\ c} \left(1, z, z^2, z^3 \right) dz, \left(i, j = x, z \right)$$

$$\left(N_{ij}^{\ a}, M_{ij}^{\ a}, P_{ij}^{\ a}, G_{ij}^{\ a} \right) = \int_{\frac{h_c}{2}}^{\frac{h_c}{2}+h_a} \sigma_{ij}^{\ a} \left(1, z, z^2, z^3 \right) dz + \int_{-\left(\frac{h_c}{2}+h_a\right)}^{-\frac{h_c}{2}} \sigma_{ij}^{\ a} \left(1, z, z^2, z^3 \right) dz, \left(i, j = x, z \right) \tag{3-23}$$

In the aforementioned equations, the superscripts "c" and "a" denote the core layer and auxetic facesheets, respectively.

3.3.2 External Work

Work is a form of energy transferred to or from the nanobeam as a result of an external force applied. In the context of the analyzed system, labor is defined as the application of energy by external forces. The initial iteration of the external labor can be articulated as follows [7]:

$$\delta V = \int_A \left[k_w \cdot \delta w_0 - kg \left(\frac{\partial^2 \delta w_0}{\partial x^2} \right) \right] dA \tag{3-24}$$

where k_w, k_P, and q denote the Winkler stiffness, shear layer stiffness, and applied transverse force, respectively. The dimensionless representation of the Winkler and Pasternak foundation is typically defined as follows:

$$K_W = \frac{k_w L^4}{D_{11}} \qquad K_g = \frac{k_g L^2}{D_{11}}$$

3.3.3 Kinetic Energy

The primary equation denoting the kinetic energy of the nanobeam can be articulated as follows:

$$\delta T = \int_v \Bigg[-I_0\left(\frac{\partial^2 u_0}{\partial t^2}\right)\delta u_0 - I_1\left(\frac{\partial^2 \phi}{\partial t^2}\right)\delta u_0 + I_3\alpha\left(\frac{\partial^2 \phi}{\partial t^2}\right)\delta u_0 + I_3\alpha\left(\frac{\partial^2 w_0}{\partial x \partial t^2}\right)$$

$$\delta u_0 - I_1\left(\frac{\partial^2 u_0}{\partial t^2}\right)\delta\phi - I_2\left(\frac{\partial^2 \phi}{\partial t^2}\right)\delta\phi + 2I_4\alpha\left(\frac{\partial^2 \phi}{\partial t^2}\right)\delta\phi + I_4\alpha\left(\frac{\partial^3 w_0}{\partial x \partial t^2}\right)$$

$$\delta\phi + I_3\alpha\left(\frac{\partial^2 u_0}{\partial t^2}\right)\delta\phi - I_6\alpha^2\left(\frac{\partial^2 \phi}{\partial t^2}\right)\delta\phi - I_6\alpha^2\left(\frac{\partial^3 w_0}{\partial x \partial t^2}\right)\delta\phi + I_6\alpha^2\left(\frac{\partial^3 \phi}{\partial x \partial t^2}\right)$$

$$\delta w_0 + I_6\alpha^2\left(\frac{\partial^4 w_0}{\partial x^2 \partial t^2}\right)\delta\phi - I_4\alpha\left(\frac{\partial^3 \phi}{\partial x \partial t^2}\right)$$

$$\delta w_0 - I_3\alpha\left(\frac{\partial^3 u_0}{\partial x \partial t^2}\right)\delta w_0 + I_O\left(\frac{\partial^2 w_0}{\partial t^2}\right)\delta w_0 \Bigg] dA \qquad (3\text{-}25)$$

where the mass moment of inertia is defined as follows:

$$I_i = \left(\int_{-h_c/2}^{h_c/2} Y_i \rho_c dz\right) + \left(\int_{\frac{(h_c+h_f)}{2}}^{\frac{-h_c}{2}} Y_i \rho_a dz + \int_{\frac{hc}{2}}^{\frac{(h_c+h_f)}{2}} Y_i \rho_a dz\right),$$

$$Y_i = \left[1, z, z^2, z^3, z^4, z^6\right], i = (1,2,3,4,6) \qquad (3\text{-}26)$$

3.3.4 Equation of Motion

To get the three equilibrium equations, it is essential to incorporate the initial variation of strain energy, the work done by applied forces, and the kinetic energy into the Euler-Lagrange equation that governs sinusoidal shear deformation. By substituting Eqs. (3–22) to (3–26) with Eq. (3–21), we obtain the following:

$$\boldsymbol{\delta u_0}: \frac{\partial N_{xx}}{\partial x} - e_{31}h_c K_c C(t)\frac{\partial^2 w_0}{\partial x \partial t} = I_0\left(\frac{\partial^2 u_0}{\partial t^2}\right) + I_1\left(\frac{\partial^2 \phi}{\partial t^2}\right)$$

$$- I_3\alpha\left(\frac{\partial^2 \phi}{\partial t^2}\right) - I_3\alpha\left(\frac{\partial^3 w_0}{\partial x \partial t^2}\right) \qquad (3\text{-}27)$$

$$\delta\phi_x: \frac{\partial M_{xx}}{\partial x} - \alpha\frac{\partial G_{xx}}{\partial x} - N_{xz} + \beta P_{xz} = I_1\left(\frac{\partial^2 u_0}{\partial t^2}\right) + I_2\left(\frac{\partial^2 \phi}{\partial t^2}\right) - 2\alpha I_4\left(\frac{\partial^2 \phi}{\partial t^2}\right) - \alpha I_4\left(\frac{\partial^3 w_0}{\partial x \partial t^2}\right) + \alpha I_3\left(\frac{\partial^2 u_0}{\partial t^2}\right) + \alpha^2 I_6\left(\frac{\partial^2 \phi}{\partial t^2}\right) + \alpha^2 I_6\left(\frac{\partial^2 u_0}{\partial t^2}\right) \quad (3\text{-}28)$$

$$\delta w_0: \alpha\frac{\partial^2 G_{xx}}{\partial x^2} + \frac{\partial N_{xz}}{\partial x} - \beta\frac{\partial P_{xz}}{\partial x} - k_w.w_0 + k_g\left(\frac{\partial^2 w_0}{\partial x^2} + \frac{\partial^2 w_0}{\partial y^2}\right) - e_{31}h_c K_c C(t)\frac{\partial^2 U_x}{\partial x \partial t} = I_0\left(\frac{\partial^2 w_0}{\partial t^2}\right) - \alpha^2 I_6\left(\frac{\partial^3 \phi}{\partial x \partial t^2}\right) - \alpha^2 I_6\left(\frac{\partial^4 w_0}{\partial x^2 \partial t^2}\right) + \alpha I_4\left(\frac{\partial^3 \phi}{\partial x \partial t^2}\right) + \alpha I_4\left(\frac{\partial^3 u_0}{\partial x \partial t^2}\right) \quad (3\text{-}29)$$

3.4 Precise Resolution

The proposed solutions must conform to the problem's boundary conditions, in alignment with Navier's method. The natural frequency can be determined by replacing the specified response (Eq. 3–30) into the governing equation. The displacements can be calculated via Navier's solution, assuming a boundary condition of simple support.

$$\begin{Bmatrix} u_0(x,y,t) \\ \phi_x(x,y,t) \\ w_0(x,y,t) \end{Bmatrix} = \begin{Bmatrix} \sum_{m=1}^{\infty}\sum_{n=1}^{\infty} U_{mn}\cos(ox)e^{i\omega t} \\ \sum_{m=1}^{\infty}\sum_{n=1}^{\infty} \phi_{xmn}\sin(ox)e^{i\omega t} \\ \sum_{m=1}^{\infty}\sum_{n=1}^{\infty} W_{mn}\sin(ox)e^{i\omega t} \end{Bmatrix} \quad (3\text{-}30)$$

where o is defined as: $o = \frac{m\pi}{L}$. By isolating the variables u_0, w_0, and ϕ in Eq. (3–27) to Eq. (3–29), the natural frequencies can be ascertained by resolving the resultant equation. To get the natural frequencies, it is essential to equate the determinant of the following equation to zero.

$$\left|[K]-\omega^2[M]+i\omega[C]\right|=0 \tag{3-31}$$

The matrices K, M, and C denote the stiffness, mass, and damping characteristics, respectively. The matrices are delineated in Appendix A.

3.5 Quantitative Findings and Discourse

This part will discuss the selected findings of the current study, concentrating on the thorough evaluation of the influence of various factors on the non-dimensional natural frequency. The thickness of the magnetostrictive layer, referred to as the core layer, is represented by the dimension h_c. The thickness of the facesheets is represented as h_a.

$$\bar{\omega}=\omega L^2\sqrt{\rho_a A / E_a I} \tag{3-32}$$

3.5.1 Findings and Analysis

Table 3.1 has been utilized to examine the initial three mode shapes of the beam to assess the influence of the auxetic inclination angle and velocity feedback gain on the natural frequencies. The data presented in Table 3.2 indicate that increasing the auxetic inclination angle results in a reduction of the natural frequencies. This may be ascribed to the reduction in flexural rigidity. This phenomenon is primarily attributable to the values of Poisson's ratio. Moreover, as illustrated in Table 3.1, increasing the feedback gain results in a reduction in the dimensionless natural frequencies of the system. This reduction enables the adjustment of the system's fluctuation frequency.

Tables 3.2 and 3.3 have been provided to examine the influence of vertical rib length and thickness of auxetic materials on natural frequencies, taking into account different values of velocity feedback gain. The results unequivocally indicate a direct association between the increase in vertical rib length and the equivalent rise in natural frequencies. The data presented in this table indicate that an increase in feedback gain correlates with a decrease in the natural frequency value.

Table 3.4 illustrates the influence of the magnetostrictive layer's thickness on the active vibrational control characteristics of the continuous

TABLE 3.1

Dimensionless Natural Frequencies of a Beam for Various Auxetic Inclination Angles and Velocity Feedback Gains for the First Three Modes (t = 1nm, l = 4nm, e = 12.5nm, L = 100nm, hc = L/10, ha = L/20)

Dimensionless Frequency	Feedback Gain	$\theta = -75°$	$\theta = -60°$	$\theta = -30°$	$\theta = -15°$	$\theta = +15°$	$\theta = +30°$	$\theta = +60°$	$\theta = +75°$
Ω_1	$k_c C(t) = 1*10^5$	24.1277	23.8162	24.689	23.7585	23.836	25.399	23.9314	24.177
Ω_1	$k_c C(t) = 5*10^5$	4.429	4.41853	4.43074	4.42458	4.42526	4.43468	4.44434	4.43123
Ω_1	$k_c C(t) = 9*10^5$	2.455	2.44983	2.45466	2.45341	2.4536	2.45533	2.4642	2.45614
Ω_2	$k_c C(t) = 1*10^5$	93.8363	92.4529	96.4339	92.1632	92.5143	99.771	92.8652	94.0529
Ω_2	$k_c C(t) = 5*10^5$	17.1404	17.0999	17.1494	17.1222	17.1251	17.1671	17.1963	17.1489
Ω_2	$k_c C(t) = 9*10^5$	9.49993	9.48031	9.49906	9.49369	9.49447	9.50208	9.53401	9.50421
Ω_3	$k_c C(t) = 1*10^5$	202.024	198.556	208.83	197.747	198.643	217.768	199.339	202.562
Ω_3	$k_c C(t) = 5*10^5$	36.662	36.5757	36.6875	36.6201	36.627	36.7318	36.7713	36.6796
Ω_3	$k_c C(t) = 9*10^5$	20.3167	20.276	20.3161	20.3032	20.3049	20.3237	20.3851	20.3254

TABLE 3.2

Dimensionless Natural Frequencies of a Beam for Various Rib Lengths and Velocity Feedback Gains (t = 0.5 nm, e = 12.5 nm, L = 100 nm, L/10, ha = L/20, θ = 60)

Natural Frequency	Feedback Gain	$l = 8nm$	$l = 4nm$	$l = 2nm$	$l = 1nm$	$l = 0.5nm$
Ω_{11}	$k_cC(t) = 1*10^5$	23.9236	23.9191	23.8572	23.7567	23.525
Ω_1	$k_cC(t) = 5*10^5$	4.42612	4.42835	4.42437	4.42097	4.40818
Ω_1	$k_cC(t) = 9*10^5$	2.45387	2.45515	2.45304	2.45136	2.4447
Ω_2	$k_cC(t) = 1*10^5$	92.9114	92.8809	92.6153	92.1706	91.1702
Ω_2	$k_cC(t) = 5*10^5$	17.1287	17.137	17.1219	17.1088	17.0601
Ω_2	$k_cC(t) = 9*10^5$	9.49553	9.50029	9.49241	9.48606	9.46096
Ω_3	$k_cC(t) = 1*10^5$	199.658	199.555	198.913	197.804	195.373
Ω_3	$k_cC(t) = 5*10^5$	36.6355	36.6521	36.6209	36.593	36.4914
Ω_3	$k_cC(t) = 9*10^5$	20.3073	20.3169	20.3008	20.2877	20.2361

TABLE 3.3

Dimensionless Natural Frequencies of a Beam for Various Auxetic Thickness Values (l = 5nm, e = 12.5nm, L = 100nm, hc = L/10, ha = L/20, θ = 45)

Natural Frequency	Feedback Gain	$t = 1nm$	$t = 0.5nm$	$t = 0.25nm$	$t = 0.1nm$	$t = 0.01nm$
Ω_1	$k_cC(t) = 1*10^5$	24.1579	24.071	24.028	24.0023	23.9869
Ω_1	$k_cC(t) = 5*10^5$	4.42762	4.42703	4.42676	4.4266	4.4265
Ω_1	$k_cC(t) = 9*10^5$	2.45414	2.45402	2.45397	2.45394	2.45393
Ω_2	$k_cC(t) = 1*10^5$	93.982	93.5841	93.3876	93.2704	93.2003
Ω_2	$k_cC(t) = 5*10^5$	17.1354	17.1328	17.1316	17.1309	17.1305
Ω_2	$k_cC(t) = 9*10^5$	9.49675	9.49622	9.496	9.49588	9.4958
Ω_3	$k_cC(t) = 1*10^5$	2002.419	201.391	200.884	200.582	200.401
Ω_3	$k_cC(t) = 5*10^5$	36.6523	36.6459	36.6428	36.641	36.6399
Ω_3	$k_cC(t) = 9*10^5$	20.3103	20.309	20.3085	20.3082	20.308

TABLE 3.4

Dimensionless Natural Frequency of the Beam at Various Auxetic Inclination Angles and Magnetostrictive Thickness (l = 5nm, e = 8nm, t = 0.5nm, L = 100nm, ha = 4nm)

Magnetostrictive Thickness	Velocity Feedback Gain	$\theta = +15°$	$\theta = +30°$	$\theta = +60°$	$\theta = +75°$
$h_m = L/5$	$k_cC(t) = 1*10^5$	30.9509	31.96	30.9318	31.0786
	$k_cC(t) = 5*10^5$	5.69844	5.70439	5.69836	5.6992
	$k_cC(t) = 9*10^5$	3.1589	3.15992	3.1589	3.15902
$h_c = L/15$	$k_cC(t) = 1*10^5$	19.2149	20.9922	19.1845	19.4238
	$k_cC(t) = 5*10^5$	3.56364	3.57367	3.56362	3.56488
	$k_cC(t) = 9*10^5$	1.97582	1.97755	1.97588	1.976
$h_c = L/25$	$k_cC(t) = 1*10^5$	13.7262	16.003	13.6916	13.9717
	$k_cC(t) = 5*10^5$	2.55344	2.56533	2.55355	2.55483
	$k_cC(t) = 9*10^5$	1.41583	1.41791	1.41598	1.41601

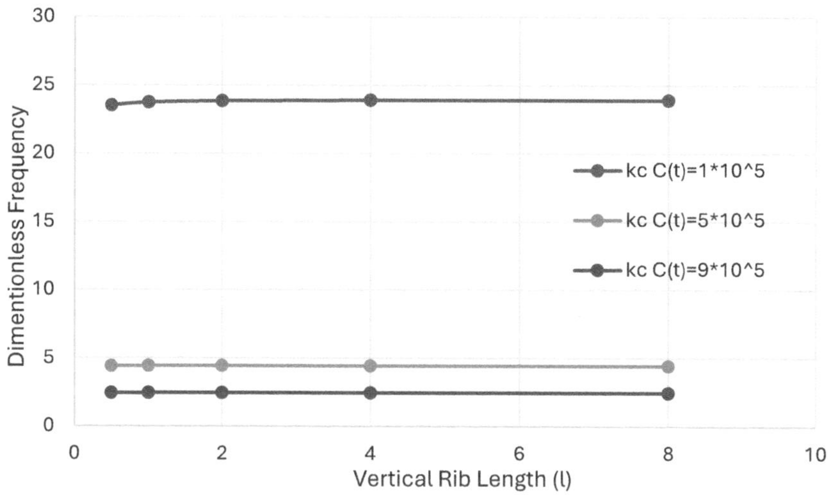

FIGURE 3.2
Dimensionless natural frequencies as a function of vertical rib length for different levels of velocity feedback gain.

system in a clear and intelligible manner. Augmenting the thickness of the magnetostrictive layer will definitely result in an increase in the non-dimensional frequencies.

Figure 3.2 illustrates that an increase in vertical rib length results in a corresponding elevation in the dimensionless natural frequencies. This phenomenon is illustrated in the graph. Moreover, augmenting the magnitude of the feedback gain facilitates a decrease in the dimensionless natural frequencies of the system. This allows us to effectively control the frequency of system variations.

Figure 3.3 clearly and succinctly illustrates the influence of the auxetic layer's thickness on the behavior of the continuous system. An increase in the core layer's thickness results in a reduction of non-dimensional frequencies, hence improving the system's stability. Moreover, it has been determined that the frequency of the structure can be altered by adjusting the wall thickness of the auxetic lattice. This was executed to attain the desired result. The beam's frequency diminishes as the ratio of its thickness to its inclined length grows. The adjustment may boost the beam's longitudinal and torsional rigidity.

Figure 3.4 is the final figure presented, illustrating the effects of two distinct levels of velocity feedback gain on the various beam components for which they are accountable. The data clearly indicate that incorporating a magnetostrictive layer into the beam elevates the dimensionless natural frequency. Furthermore, of the three kinds, the pure magnetostrictive beam exhibits the highest natural frequency.

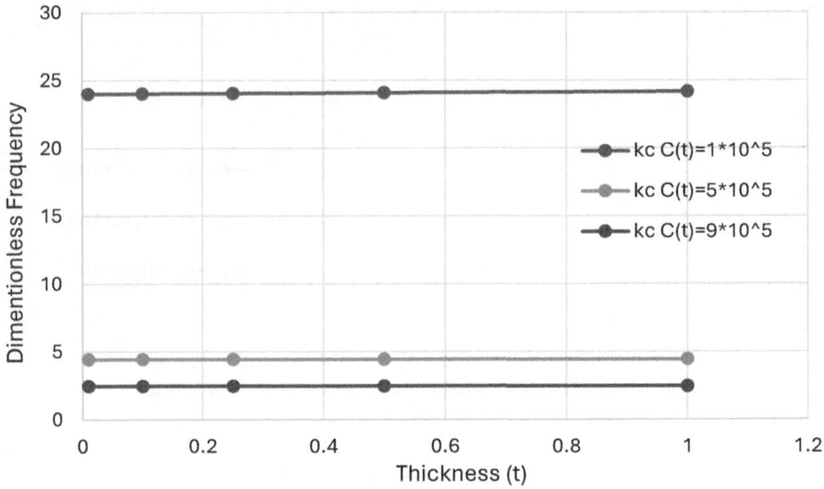

FIGURE 3.3
Dimensionless natural frequencies as a function of thickness (t) at different levels of velocity feedback gain.

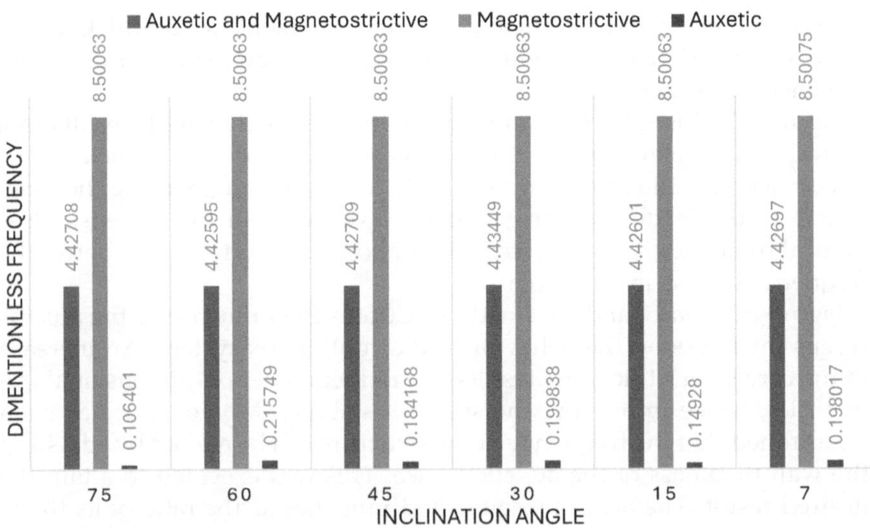

FIGURE 3.4
Comparison of dimensionless natural frequencies for three distinct beam types: (a) velocity feedback gain = 5 × 10^5, (b) velocity feedback gain = 9 × 10^5.

3.5.2 Final Assessment

This chapter seeks to examine the vibrational characteristics of a composite magnetostrictive nanobeam integrated with auxetic facesheets. The inquiry utilizes the higher-order shear deformation beam theory. Moreover, the

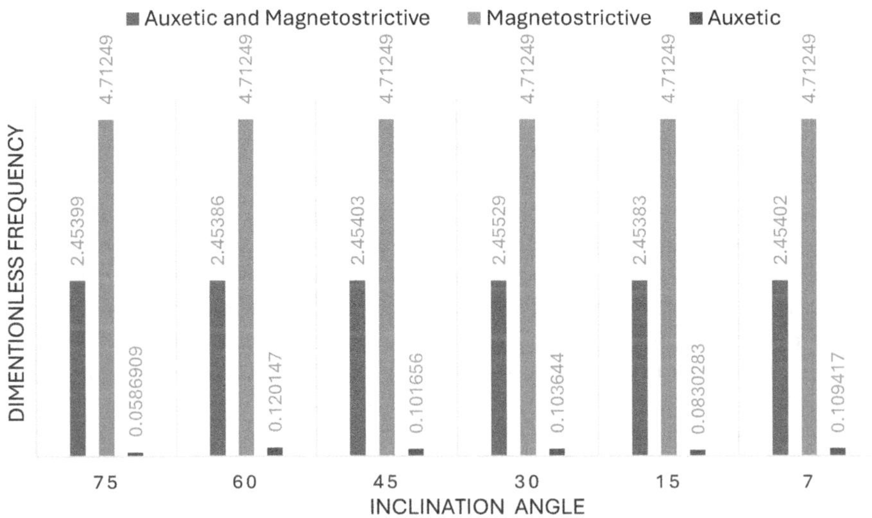

FIGURE 3.4
(Continued)

Hamiltonian framework is utilized to derive the governing equations of the system. The following are significant results of the present study:

- The investigation revealed that of the three beam types analyzed, the magnetostrictive plate exhibited the highest dimensionless natural frequency values.
- The dimensionless natural frequencies decrease as the inclination angle increases.
- The dimensionless natural frequency rises with an increase in rib length.
- Magnetostrictive material regulates the behavior of the beam.

3.6 Vibration of Active Material Plates

Vibration research of active material plates is one of the most important frontiers in modern engineering, integrating structural dynamics, materials science, and control of smart systems. Unlike passive plates, the structure is bonded with materials like piezoelectric ceramics, shape memory alloys, or magnetostrictive compounds that have sensor as well as actuator capabilities. This unique feature enables real-time monitoring, control, and suppression of vibrational responses. Thus, their behavior is complex, involving coupled electromechanical or thermomechanical interactions, and forms the core of the development of aerospace technologies, robotics, precision manufacturing, and adaptive structures.

3.7 Background

The present section seeks to regulate the vibrational characteristics of an auxetic plate coated with magnetostrictive material. The kinematic relationships of the plate, supported by a Winkler-Pasternak medium, are articulated according to the first-order shear deformation theory. The governing equations are formulated using Hamilton's principle and solved analytically by Navier's method. The influence of parameters, including auxetic inclination angle, auxetic rib length, and feedback gain, on the system's control behavior is meticulously observed. The findings demonstrate that incorporating an auxetic core into a magnetostrictive plate enhances the dimensionless natural frequency. The findings from this study may enhance the development of applications like sensor design, actuator improvement, and vibration mitigation systems. The authors contend that the synergistic impacts of tunable qualities in magnetostrictive and auxetic materials on the frequency characteristics of a smart nanoplate, intended to manipulate its vibrational behavior, remain insufficiently investigated. This study seeks to examine the vibration control of plates featuring an auxetic core and coated magnetostrictive materials, thereby addressing a deficiency in the current literature.

3.8 Formulation

This project aims to investigate plates with auxetic cores and coated magnetostrictive materials for vibration control. The coordinates (x, y,

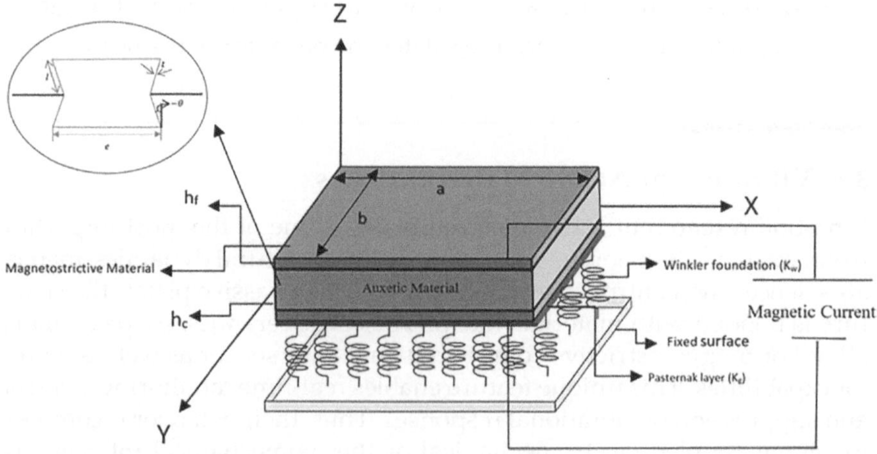

FIGURE 3.5
Architectural plan of the edifice.

and z) are established inside a Cartesian coordinate system, with the origin positioned at one corner of the plate, as depicted in Figure 3.5. The Winkler-Pasternak foundation has been employed to consider the foundation's influence. The FSDT is utilized to define the motion of the composite plate.

3.8.1 Kinematic Equations

The FSDT will be employed to delineate the issue in this essay. The corresponding FSDT displacement field can be articulated as [9, 10]:

$$U_x(x,y,z,t) = u_0(x,y,t) + \phi_x(x,y,t) \tag{3-33}$$

$$V_y(x,y,z,t) = v_0(x,y,t) + \phi_y(x,y,t) \tag{3-34}$$

$$W_z(x,y,z,t) = w_0(x,y,t) \tag{3-35}$$

The comprehensive displacement fields along the Cartesian x, y, and z axes are represented by U_x, V_y, and W_z.

3.8.2 Auxetic Materials

The auxetic core layer consists of cells organized in a honeycomb configuration. The geometrical characteristics of the individual unit cell are believed to dictate the mechanical properties of the honeycomb auxetic core layer, encompassing the Young's modulus, shear modulus, negative Poisson's ratio, and thermal expansion coefficients.

3.8.2.1 Auxetic Core

The prospective results for the auxetic core's effective mechanical properties are as Eq. (3–4) to Eq. (3–8).

3.8.3 Stress-Strain Relationship of Auxetic Materials

For auxetic materials, the correlation between stress and strain can be expressed as [11, 12]:

$$\sigma_{xx}^a = C_{11}^a \varepsilon_{xx} + C_{12}^a \varepsilon_{yy} \tag{3-36}$$

$$\sigma_{yy}^a = C_{21}^a \varepsilon_{xx} + C_{22}^a \varepsilon_{yy} \tag{3-37}$$

$$\tau_{yz}^a = C_{44}^a \gamma_{yz} \tag{3-38}$$

$$\tau_{xz}^{a} = C_{55}^{a}\gamma_{xz} \tag{3-39}$$

$$\tau_{xy}^{a} = C_{66}^{a}\gamma_{xy} \tag{3-40}$$

Auxetic layer non-zero stresses are characterized as follows:

$$\begin{Bmatrix} \varepsilon_{xx}^{a} \\ \varepsilon_{yy}^{a} \\ \varepsilon_{yz}^{a} \\ \varepsilon_{xz}^{a} \\ \varepsilon_{xy}^{a} \end{Bmatrix} = \begin{Bmatrix} \varepsilon_{xx}^{a} \\ \varepsilon_{yy}^{a} \\ \varepsilon_{yz}^{a} \\ \varepsilon_{xz}^{a} \\ \varepsilon_{xy}^{a} \end{Bmatrix} + z \begin{Bmatrix} \kappa_{xx}^{a} \\ \kappa_{yy}^{a} \\ \kappa_{yz}^{a} \\ \kappa_{xz}^{a} \\ \kappa_{xy}^{a} \end{Bmatrix} \tag{3-41}$$

in which,

$$\begin{Bmatrix} \varepsilon_{xx}^{a} \\ \varepsilon_{yy}^{a} \\ \varepsilon_{yz}^{a} \\ \varepsilon_{xz}^{a} \\ \varepsilon_{xy}^{a} \end{Bmatrix} = \begin{Bmatrix} \dfrac{\partial u_0}{\partial x} \\ \dfrac{\partial v_0}{\partial y} \\ \left(\phi_y + \dfrac{\partial w_0}{\partial y}\right) \\ \left(\phi_x + \dfrac{\partial w_0}{\partial x}\right) \\ \left(\dfrac{\partial u_0}{\partial y} + \dfrac{\partial v_0}{\partial x}\right) \end{Bmatrix} \begin{Bmatrix} \kappa_{xx}^{a} \\ \kappa_{yy}^{a} \\ \kappa_{yz}^{a} \\ \kappa_{xz}^{a} \\ \kappa_{xy}^{a} \end{Bmatrix} = \begin{Bmatrix} \dfrac{\partial \phi_x}{\partial x} \\ \dfrac{\partial \phi_y}{\partial y} \\ 0 \\ 0 \\ \dfrac{\partial \phi_x}{\partial y} + \dfrac{\partial \phi_y}{\partial x} \end{Bmatrix} \tag{3-42}$$

3.8.4 Magnetostrictive Substances

The Joule effect, or magnetostriction, refers to the alteration of a ferromagnetic material in response to an external magnetic field. The ramifications of this phenomenon can be observed in numerous applications. In 1842, English scientist James Joule was the first to elucidate the phenomenon known as magnetostriction. He discovered that a control sample composed of ferromagnetic material, such as iron, exhibits a length change when situated near a magnetic field.

3.8.4.1 Terfenol-D

Terfenol-D is the preeminent magnetostrictive material, exhibiting optimal characteristics for strain and Curie temperature. The research and advancement of magnetostrictive supermaterials commenced in the 1960s by scholars, including Clark et al. Terfenol-D (refer to Figure 3.6) was discovered in the 1970s at the Naval Ordnance Laboratory by a research team under Keller's leadership. Keller was accountable for the laboratory's designation.

3.8.5 Stress-Strain Relationship for Magnetostrictive Materials

The formula denotes the stress, strain, and magnetic susceptibility of the magnetostrictive layer.

$$\sigma_{xx}^{m} = C_{11}^{m}\varepsilon_{xx} + C_{12}^{m}\varepsilon_{yy} - e_{31}H_{zz} \tag{3-43}$$

$$\sigma_{yy}^{m} = C_{21}^{m}\varepsilon_{xx} + C_{22}^{m}\varepsilon_{yy} - e_{32}H_{zz} \tag{3-44}$$

$$\tau_{yz}^{m} = C_{44}^{m}\gamma_{yz} \tag{3-45}$$

$$\tau_{xz}^{m} = C_{55}^{m}\gamma_{xz} \tag{3-46}$$

$$\tau_{xy}^{m} = C_{66}^{m}\gamma_{xy} \tag{3-47}$$

where stress and strain tensors are denoted by arrays of the format $\sigma_{ij}, \varepsilon_{ij}\,(i, j = x, y, z)$. Additionally, the magnetic field is denoted by H_{zz}. Furthermore, note that C_{ij} denotes the arrays of the elasticity tensor. Assuming zero thickness stretching, or $\varepsilon_{zz} = 0$, the elastic constants in the plane of stress can be expressed as follows:

$$C_{11}^{m} = \frac{E_m}{1-\vartheta_m^{\,2}} \quad C_{12}^{m} = \frac{\vartheta_m E_m}{1-\vartheta_m^{\,2}} \quad C_{66}^{m} = \frac{E_m}{2(1+\vartheta_m)} \quad C_{55}^{m} = C_{44}^{m} = k_s C_{66}^{m} \tag{3-48}$$

a) integrated b) powdered

FIGURE 3.6
Commercial forms of Terfenol-D: (a) integrated and (b) powdered.

E_m and v_m represent the elasticity modulus and Poisson's ratio of the magnetostrictive layers, respectively. The shear correction parameter, denoted as the constant k_s, has been attributed varying values by different theories.

3.8.6 Velocity Feedback Gain

The magnetic field (H_{zz}) can be computed using the equation presented in Eq. (3–13).

3.9 Governing Equation

This section derives the motion equations using the energy technique. The concept of Hamilton will be employed for this aim. A potential articulation of this concept is presented in Eq. (3–21).

The initial fluctuation of strain energy can be expressed as follows:

$$\begin{aligned}\delta U^a = \int_0^a\int_0^b & N_{xx}^a\left(\delta\varepsilon_{xx}^a\right)+M_{xx}^a\left(\delta\kappa_{xx}^a\right)+N_{yy}^a\left(\delta\varepsilon_{yy}^a\right)+M_{yy}^a\left(\delta\kappa_{yy}^a\right)\\ &+N_{xz}^a\left(\delta\varepsilon_{xz}^a\right)+M_{xz}^a\left(\delta\kappa_{xz}^a\right)+N_{yz}^a\left(\delta\varepsilon_{yz}^a\right)+M_{yz}^a\left(\delta\kappa_{yz}^a\right)\\ &+N_{xy}^a\left(\delta\varepsilon_{xy}^a\right)+M_{xy}^a\left(\delta\kappa_{xy}^a\right)dA\end{aligned} \quad (3\text{-}49)$$

$$\begin{aligned}\delta U^m = \int_0^a\int_0^b & N_{xx}^m\left(\delta\varepsilon_{xx}^m\right)+M_{xx}^m\left(\delta\kappa_{xx}^m\right)+N_{yy}^m\left(\delta\varepsilon_{yy}^m\right)+M_{yy}^m\left(\delta\kappa_{yy}^m\right)\\ &+N_{xz}^m\left(\delta\varepsilon_{xz}^m\right)+M_{xz}^m\left(\delta\kappa_{xz}^m\right)+N_{yz}^m\left(\delta\varepsilon_{yz}^m\right)+M_{yz}^m\left(\delta\kappa_{yz}^m\right)\\ &+N_{xy}^m\left(\delta\varepsilon_{xy}^m\right)+M_{xy}^m\left(\delta\kappa_{xy}^m\right)dA\end{aligned} \quad (3\text{-}50)$$

in which,

$$\left(N_{ij}{}^a, M_{ij}{}^a\right)=\int_{-\frac{h_a}{2}}^{\frac{h_a}{2}}\sigma_{ij}{}^a(1,z)dz,\ (i,j=x,y,z) \quad (3\text{-}51)$$

$$\left(N_{ij}{}^m, M_{ij}{}^m\right)=\int_{\frac{h_a}{2}}^{\frac{h_a}{2}+h_m}\sigma_{ij}{}^m(1,z)dz+\int_{-\left(\frac{h_a}{2}+h_m\right)}^{-\frac{h_a}{2}}\sigma_{ij}{}^m(1,z)dz,\ (i,j=x,y,z) \quad (3\text{-}52)$$

The auxetic and magnetostrictive layers are represented by the superscripts "a" and "m" in the previously discussed equations. Furthermore, the forces $N_{i.j}$ and the moments $M_{i,j}$ are specified. The correlations between forces and bending moments are as follows:

$$\begin{bmatrix} N_{xx} \\ N_{yy} \\ N_{yz} \\ N_{xz} \\ N_{xy} \end{bmatrix} = \left(\begin{bmatrix} A_{11} & A_{12} & 0 & 0 & 0 \\ A_{21} & A_{22} & 0 & 0 & 0 \\ 0 & 0 & A_{44} & 0 & 0 \\ 0 & 0 & 0 & A_{55} & 0 \\ 0 & 0 & 0 & 0 & A_{66} \end{bmatrix} \begin{bmatrix} \dfrac{\partial u_0}{\partial x} \\ \dfrac{\partial v_0}{\partial y} \\ \phi_y + \dfrac{\partial w_0}{\partial y} \\ \phi_x + \dfrac{\partial w_0}{\partial x} \\ \dfrac{\partial u_0}{\partial y} + \dfrac{\partial v_0}{\partial x} \end{bmatrix} + \begin{bmatrix} B_{11} & B_{12} & 0 & 0 & 0 \\ B_{21} & B_{22} & 0 & 0 & 0 \\ 0 & 0 & B_{44} & 0 & 0 \\ 0 & 0 & 0 & B_{55} & 0 \\ 0 & 0 & 0 & 0 & B_{66} \end{bmatrix} \begin{bmatrix} \dfrac{\partial \phi_x}{\partial x} \\ \dfrac{\partial \phi_y}{\partial y} \\ 0 \\ 0 \\ \dfrac{\partial \phi_x}{\partial y} + \dfrac{\partial \phi_y}{\partial x} \end{bmatrix} \right) \tag{3-53}$$

$$\begin{bmatrix} M_{xx} \\ M_{yy} \\ M_{yz} \\ M_{xz} \\ M_{xy} \end{bmatrix} = \left(\begin{bmatrix} B_{11} & B_{12} & 0 & 0 & 0 \\ B_{21} & B_{22} & 0 & 0 & 0 \\ 0 & 0 & B_{44} & 0 & 0 \\ 0 & 0 & 0 & B_{55} & 0 \\ 0 & 0 & 0 & 0 & B_{66} \end{bmatrix} \begin{bmatrix} \dfrac{\partial u_0}{\partial x} \\ \dfrac{\partial v_0}{\partial y} \\ \phi_y + \dfrac{\partial w_0}{\partial y} \\ \phi_x + \dfrac{\partial w_0}{\partial x} \\ \dfrac{\partial u_0}{\partial y} + \dfrac{\partial v_0}{\partial x} \end{bmatrix} + \begin{bmatrix} D_{11} & D_{12} & 0 & 0 & 0 \\ D_{21} & D_{22} & 0 & 0 & 0 \\ 0 & 0 & D_{44} & 0 & 0 \\ 0 & 0 & 0 & D_{55} & 0 \\ 0 & 0 & 0 & 0 & D_{66} \end{bmatrix} \begin{bmatrix} \dfrac{\partial \phi_x}{\partial x} \\ \dfrac{\partial \phi_y}{\partial y} \\ 0 \\ 0 \\ \dfrac{\partial \phi_x}{\partial y} + \dfrac{\partial \phi_y}{\partial x} \end{bmatrix} \right) \tag{3-54}$$

in which,

$$\begin{bmatrix} A_{11} & B_{11} & D_{11} \\ A_{22} & B_{22} & D_{22} \\ A_{12} & B_{12} & D_{12} \\ A_{44} & B_{44} & D_{44} \\ A_{55} & B_{55} & D_{55} \\ A_{66} & B_{66} & D_{66} \end{bmatrix} = \int_{-\frac{h_a}{2}}^{\frac{h_a}{2}} \begin{bmatrix} 1 & z & z^2 \end{bmatrix} \begin{bmatrix} C_{11}^a \\ C_{22}^a \\ C_{12}^a \\ C_{44}^a \\ C_{55}^a \\ C_{66}^a \end{bmatrix} dz + \int_{\frac{h_a}{2}}^{\frac{h_a}{2}+h_m} \begin{bmatrix} 1 & z & z^2 \end{bmatrix} \begin{bmatrix} C_{11}^m \\ C_{22}^m \\ C_{12}^m \\ C_{44}^m \\ C_{55}^m \\ C_{66}^m \end{bmatrix} dz + \int_{-\left(\frac{h_a}{2}+h_m\right)}^{-\frac{h_a}{2}} \begin{bmatrix} 1 & z & z^2 \end{bmatrix} \begin{bmatrix} C_{11}^m \\ C_{22}^m \\ C_{12}^m \\ C_{44}^m \\ C_{55}^m \\ C_{66}^m \end{bmatrix} dz \tag{3-55}$$

The initial version of the exterior work can be articulated as [7]:

$$\delta V = k_w.w - k_g\left(\frac{\partial^2 w}{\partial x^2} + \frac{\partial^2 w}{\partial y^2}\right) \tag{3-56}$$

where k_w denotes the Winkler stiffness and k_g represents the shear layer stiffness, respectively.

The initial fluctuation of kinetic energy can be expressed as follows:

$$\delta T = \frac{1}{2}\iint\left[I_0\left(\frac{\partial^2 u_0}{\partial t^2}\right)\delta u_0 + I_1\left(\frac{\partial^2 u_0}{\partial t^2}\right)\delta\phi_x + I_1\left(\frac{\partial^2 \phi_x}{\partial t^2}\right)\delta u_0 \right.$$
$$+ I_2\left(\frac{\partial^2 \phi_x}{\partial t^2}\right)\delta\phi_x + I_0\left(\frac{\partial^2 v_0}{\partial t^2}\right)\delta v_0 + I_1\left(\frac{\partial^2 v_0}{\partial t^2}\right)\delta\phi_y$$
$$\left. + I_1\left(\frac{\partial^2 \phi_y}{\partial t^2}\right)\delta v_0 + I_2\left(\frac{\partial^2 \phi_y}{\partial t^2}\right)\delta\phi_y + I_0\left(\frac{\partial^2 w_0}{\partial t^2}\right)\delta w_0 \right] dA \quad (3\text{-}57)$$

The mass moments of inertia are defined as follows:

$$I_0 = \int_{-\frac{h_a}{2}}^{\frac{h_a}{2}} \rho_a dz + \left(\int_{-\frac{(h_a+h_m)}{2}}^{\frac{-h_a}{2}} \rho_m(z)dz + \int_{\frac{h_a}{2}}^{\frac{(h_a+h_m)}{2}} \rho_m(z)dz \right) \quad (3\text{-}58)$$

$$I_1 = \int_{-h_a/2}^{h_a/2} z\rho_a dz + \left(\int_{-\frac{(h_a+h_m)}{2}}^{\frac{-h_a}{2}} z\rho_m(z)dz + \int_{\frac{h_a}{2}}^{\frac{(h_a+h_m)}{2}} z\rho_m(z)dz \right) \quad (3\text{-}59)$$

$$I_2 = \int_{-h_a/2}^{h_a/2} z^2\rho_a dz + \left(\int_{-\frac{(h_a+h_m)}{2}}^{\frac{-h_a}{2}} z^2\rho_m(z)dz + \int_{\frac{h_a}{2}}^{\frac{(h_a+h_m)}{2}} z^2\rho_m(z)dz \right) \quad (3\text{-}60)$$

When Eqs (3–49) to (3–60) are replaced with Eq. (3–13), governing equations are as follows:

$$\delta u_0: \frac{\partial N_{xx}}{\partial x} + \frac{\partial N_{xy}}{\partial y} - e_{31}h_m K_c C(t)\frac{\partial^2 w_0}{\partial x \partial t} = I_0\frac{\partial^2 u}{\partial t^2} + I_1\frac{\partial^2 \phi_x}{\partial t^2} \quad (3\text{-}61)$$

$$\delta v_0: \frac{\partial N_{xx}}{\partial y} + \frac{\partial N_{xy}}{\partial x} - e_{32}h_m K_c C(t)\frac{\partial^2 w_0}{\partial y \partial t} = I_0\frac{\partial^2 v}{\partial t^2} + I_1\frac{\partial^2 \phi_y}{\partial t^2} \quad (3\text{-}62)$$

$$\delta w_0: \frac{\partial N_{xz}}{\partial x} + \frac{\partial N_{yz}}{\partial y} - k_w.w + k_g\left(\frac{\partial^2 w}{\partial x^2} + \frac{\partial^2 w}{\partial y^2}\right) - e_{31}h_m K_c C(t)\frac{\partial^2 U_x}{\partial x \partial t}$$
$$- e_{32}h_m K_c C(t)\frac{\partial^2 V_y}{\partial y \partial t} = I_0 \frac{\partial^2 w}{\partial t^2} \qquad (3\text{-}63)$$

$$\delta\varnothing_x: \frac{\partial M_{xx}}{\partial x} + \frac{\partial M_{xy}}{\partial y} - N_{xz} = I_1 \frac{\partial^2 u}{\partial t^2} + I_2 \frac{\partial^2 \phi_x}{\partial t^2} \qquad (3\text{-}64)$$

$$\delta\varnothing_y: \frac{\partial M_{yy}}{\partial y} + \frac{\partial M_{xy}}{\partial x} - N_{yz} = I_1 \frac{\partial^2 v}{\partial t^2} + I_2 \frac{\partial^2 \phi_y}{\partial t^2} \qquad (3\text{-}65)$$

3.10 Analytical Resolution

An analytical solution formulated by Navier was employed to resolve the governing equation. The proposed solutions to Navier's problem must adhere to the boundary conditions for the solution to be valid. To proceed, the natural frequency can be determined by substituting the proposed response into the Eq. (3–66), which is the equation that regulates the system. The Navier technique has been effectively employed to address many issues in vibration analysis, proving to be a definitive method for solving partial differential equations with excellent accuracy, convergence, and performance. Furthermore, the technique has been effectively employed to resolve various issues in the domain of vibration analysis.

Assuming simply supported boundary conditions, the displacements can be defined according to Navier's solution as follows:

$$\begin{Bmatrix} u_0(x,y,t) \\ v_0(x,y,t) \\ w_0(x,y,t) \\ \phi_x(x,y,t) \\ \phi_y(x,y,t) \end{Bmatrix} = \begin{Bmatrix} \sum_{m=1}^{\infty}\sum_{n=1}^{\infty} U_{mn}\cos(\alpha x)\sin(\beta y)e^{i\omega t} \\ \sum_{m=1}^{\infty}\sum_{n=1}^{\infty} V_{mn}\sin(\alpha x)\cos(\beta y)e^{i\omega t} \\ \sum_{m=1}^{\infty}\sum_{n=1}^{\infty} W_{mn}\sin(\alpha x)\sin(\beta y)e^{i\omega t} \\ \sum_{m=1}^{\infty}\sum_{n=1}^{\infty} \phi_{xmn}\cos(\alpha x)\sin(\beta y)e^{i\omega t} \\ \sum_{m=1}^{\infty}\sum_{n=1}^{\infty} \phi_{ymn}\sin(\alpha x)\cos(\beta y)e^{i\omega t} \end{Bmatrix} \qquad (3\text{-}66)$$

where U_{mn}, V_{mn}, W_{mn}, ϕ_{xmn}, and ϕ_{ymn} represent the amplitudes of the vibration. Additionally, ω denotes the fundamental frequency. Furthermore, α and β are defined as $\beta = n\pi/y$ and $\alpha = m\pi/a$.

The natural frequencies can be determined by substituting Eq. (3–66) for equations. Solving for $u_0, v_0, w_0, \phi_x, \phi_y,$ in the range of Eqs. (3–61) to (3–65). The determinant of the following equation must be equated to zero to ascertain the fundamental frequencies.

$$\left|[K]-\omega^2[M]+i\omega[C]\right|=0 \tag{3-67}$$

K, M, and C represent the stiffness, mass, and damping matrices, respectively, as delineated in Appendix B.

3.10.1 Findings and Analysis

Table 3.5 has been created to monitor the initial three-mode forms of the plate to assess the influence of the auxetic inclination angle and velocity feedback gain on the natural frequencies. The table indicates that an increase in the auxetic inclination angle leads to a drop in natural frequencies, attributable to a decrease in flexural stiffness. This occurrence may primarily result from the values of Poisson's ratio. Furthermore, an increase in the feedback gain value results in a reduction in the dimensionless natural frequencies of the system, hence facilitating control over the system's frequency fluctuations.

Tables 3.6 and 3.7 are presented to analyze the impact of vertical rib length and thickness of auxetic materials on natural frequencies for various values of velocity feedback gain. The data indicate that an increase in the vertical rib length correlates with an increase in natural frequencies. This table demonstrates that an increase in the feedback gain correlates with a decline in the natural frequency value.

Table 3.8 demonstrates the impact of the magnetostrictive layer thickness on the active vibrational control behavior of the continuous system. Clearly, an increase in the thickness of the magnetostrictive layer results in a decrease in non-dimensional frequencies, hence enhancing system stability.

Figure 3.7 clearly illustrates that an increase in the inclination angle results in a drop in the dimensionless natural frequencies, a phenomenon previously explained from a physical perspective.

Furthermore, the figure illustrates the impact of the feedback gain parameter on the dimensionless natural frequency and demonstrates that an increase in the feedback gain results in a substantial decrease in natural frequencies, attributable to the dampening influence of the magnetostriction phenomena on the system's dynamics. Velocity feedback gain plays a crucial function in the stability of the system.

Figure 3.8 examines the impact of vertical rib length on system stability. The figure demonstrates that an increase in the vertical rib length results

TABLE 3.5

Dimensionless Natural Frequencies of a Square Plate for Various Auxetic Inclination Angles and Velocity Feedback Gains for the First Three Modes (t = 1mm, l = 4mm, e = 8mm, a = b = 60mm, hc = 4mm, hf = 0.1mm)

Dimensionless Frequency	Feedback Gain	$\theta = -75°$	$\theta = -60°$	$\theta = -30°$	$\theta = -15°$	$\theta = +15°$	$\theta = +30°$	$\theta = +60°$	$\theta = +75°$
Ω_{11}	$k_c C(t) = 0$	2.8911	2.88791	2.88325	2.88314	2.88416	2.88518	2.88432	2.88343
Ω_{11}	$k_c C(t) = 1*10^5$	2.8883	2.88511	2.88046	2.88034	2.88136	2.88238	2.88153	2.88063
Ω_{11}	$k_c C(t) = 5*10^5$	2.87713	2.87395	2.8693	2.86919	2.87021	2.87123	2.87037	2.86948
Ω_{11}	$k_c C(t) = 9*10^5$	2.866	2.86283	2.85819	2.85808	2.8591	2.86012	2.85927	2.85836
Ω_{22}	$k_c C(t) = 0$	10.4732	10.4518	10.4263	10.4254	10.4283	10.4323	10.4279	10.428
Ω_{22}	$k_c C(t) = 1*10^5$	10.4639	10.4426	10.4171	10.4162	10.4191	10.4231	10.4187	10.4189
Ω_{22}	$k_c C(t) = 5*10^5$	10.4271	10.4058	10.3805	10.3795	10.3824	10.3864	10.382	10.3821
Ω_{22}	$k_c C(t) = 9*10^5$	10.3903	10.3692	10.3439	10.3429	10.3459	10.3498	10.3456	10.3456
Ω_{33}	$k_c C(t) = 0$	20.7403	20.6778	20.6095	20.6063	20.6105	20.619	20.6077	20.6151
Ω_{33}	$k_c C(t) = 1*10^5$	20.724	20.6616	20.5933	20.5902	20.5944	20.6028	20.5916	20.5989
Ω_{33}	$k_c C(t) = 5*10^5$	20.6589	20.5967	20.5287	20.5256	20.5298	20.5383	20.5271	20.5343
Ω_{33}	$k_c C(t) = 9*10^5$	20.594	20.5321	20.4644	20.4613	20.4655	20.474	20.4628	20.4699

TABLE 3.6

Dimensionless Natural Frequencies of a Square Plate with Varying Rib Lengths and Velocity Feedback Gains (t = 0.25 mm, e = 8 mm, a = b = 60 mm, hc = 4 mm, hf = 0.1 mm, θ = 60°)

Natural Frequency	Feedback Gain	$l = 8$mm	$l = 4$mm	$l = 2$mm	$l = 1$mm	$l = 0.5$mm
Ω_{11}	$k_c C(t) = 0*10^5$	2.88417	2.88418	2.88381	2.88315	2.87827
Ω_{11}	$k_c C(t) = 1*10^5$	2.88138	2.88138	2.88102	2.88035	2.87547
Ω_{11}	$k_c C(t) = 5*10^5$	2.87022	2.87023	2.86986	2.8692	2.86432
Ω_{12}	$k_c C(t) = 0*10^5$	6.83553	6.83553	6.83475	6.83343	6.82466
Ω_{12}	$k_c C(t) = 1*10^5$	6.82923	6.82924	6.82845	6.82714	6.81836
Ω_{12}	$k_c C(t) = 5*10^5$	6.80411	6.80411	6.80332	6.80201	6.79324
Ω_{22}	$k_c C(t) = 0*10^5$	10.4291	10.4291	10.4279	10.4253	10.4079
Ω_{22}	$k_c C(t) = 1*10^5$	10.4199	10.4199	10.4187	10.4161	10.3987
Ω_{22}	$k_c C(t) = 5*10^5$	10.3832	10.3832	10.382	10.3794	10.362

TABLE 3.7

Dimensionless Natural Frequencies of a Square Plate for Various Auxetic Thickness Values (l = 2mm, e = 8mm, a = b = 60mm, hc = 4mm, hf = 0.1mm, θ = 60°)

Natural Frequency	Feedback Gain	$t = 1$mm	$t = 0.5$mm	$t = 0.25$mm	$t = 0.1$mm	$t = 0.01$mm
Ω_{11}	$k_c C(t) = 0*10^5$	2.86089	2.88128	2.88381	2.88415	2.88417
Ω_{11}	$k_c C(t) = 1*10^5$	2.85809	2.87848	2.88102	2.88136	2.88138
Ω_{11}	$k_c C(t) = 5*10^5$	2.84694	2.86732	2.86986	2.8702	2.87023
Ω_{12}	$k_c C(t) = 0*10^5$	6.78499	6.82924	6.83475	6.83548	6.83553
Ω_{12}	$k_c C(t) = 1*10^5$	6.77869	6.82294	6.82845	6.82919	6.82924
Ω_{12}	$k_c C(t) = 5*10^5$	6.75335	6.79781	6.80332	6.80406	6.80411
Ω_{22}	$k_c C(t) = 0*10^5$	10.3491	10.4191	10.4279	10.429	10.4291
Ω_{22}	$k_c C(t) = 1*10^5$	10.3399	10.41	10.4187	10.4198	10.4199
Ω_{12}	$k_c C(t) = 5*10^5$	10.3032	10.3733	10.382	10.3831	10.3832

TABLE 3.8

Dimensionless Natural Frequency of the Plate for Various Auxetic Inclination Angles and Magnetostrictive Thickness (l = 2mm, e = 8mm, t = 0.25mm, a = b = 60mm, ha = 4mm)

Magnetostrictive Thickness	Velocity Feedback Gain	$\theta = +15°$	$\theta = +30°$	$\theta = +60°$	$\theta = +75°$
$h_m = 0.15$mm	$k_c C(t) = 0*10^5$	3.4475	3.44758	3.44721	3.44761
	$k_c C(t) = 1*10^5$	3.44351	3.44359	3.44321	3.44362
	$k_c C(t) = 5*10^5$	3.42758	3.42766	3.42728	3.42769
$h_c = 0.1$mm	$k_c C(t) = 0*10^5$	2.88419	2.8843	2.88381	2.88433
	$k_c C(t) = 1*10^5$	2.8814	2.8815	2.88102	2.88154
	$k_c C(t) = 5*10^5$	2.87024	2.87035	2.86986	2.87038
$h_c = 0.01$mm	$k_c C(t) = 0*10^5$	0.959487	0.959846	0.958167	0.959975
	$k_c C(t) = 1*10^5$	0.959177	0.959537	0.957858	0.959666
	$k_c C(t) = 5*10^5$	0.957941	0.9583	0.956621	0.95843

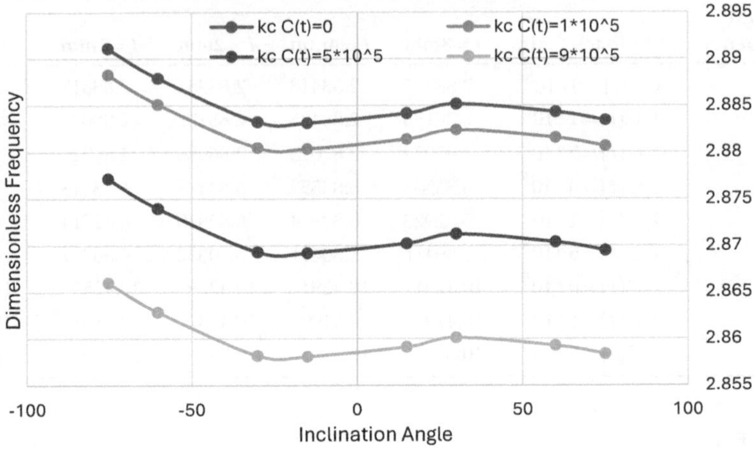

FIGURE 3.7
Dimensionless natural frequencies as a function of inclination angle for different levels of velocity feedback gain (t = 1mm, l = 4mm, e = 8mm, a = b = 60mm, hc = 4mm, hf = 0.1mm).

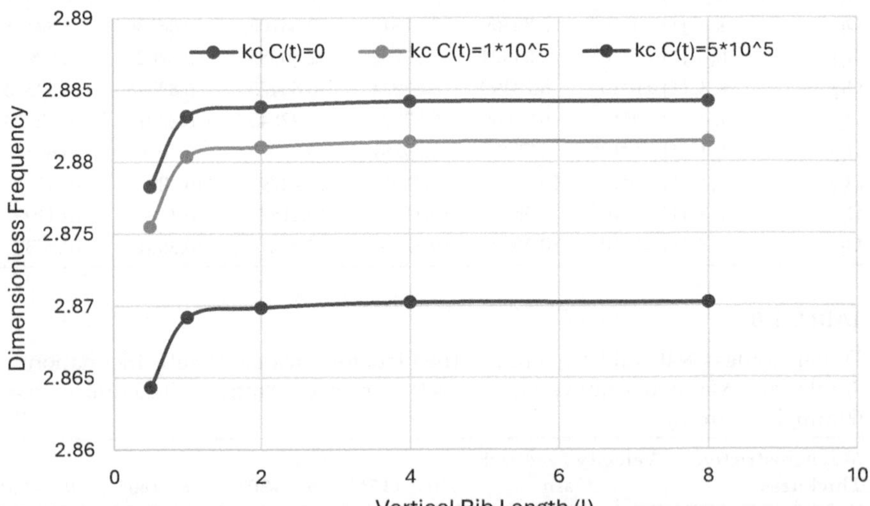

FIGURE 3.8
Effect of vertical rib length on system stability.

in a corresponding augmentation of the dimensionless natural frequencies. Moreover, augmenting the feedback gain diminishes the dimensionless natural frequencies of the system, hence facilitating the regulation of the system's fluctuation frequency.

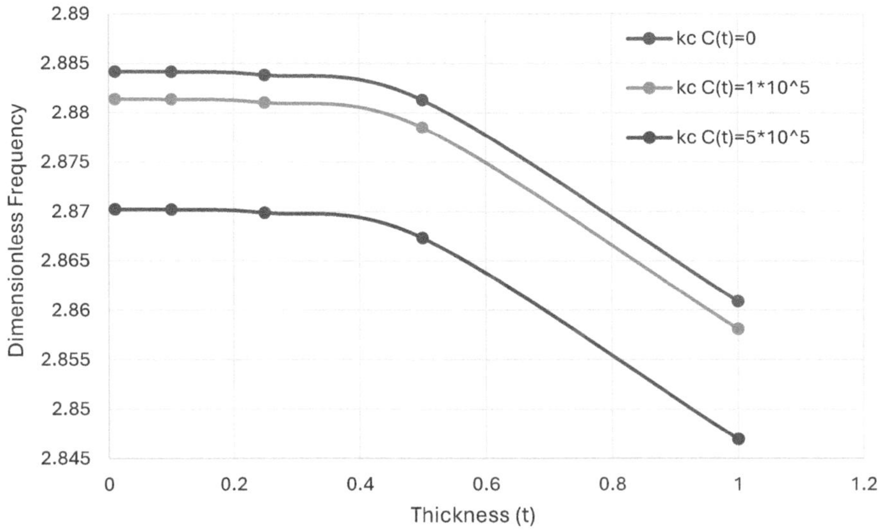

FIGURE 3.9
Effect of the auxetic layer's thickness on system stability.

Figure 3.9 clearly illustrates the impact of the auxetic layer's thickness on the behavior of the continuous system. Clearly, an increase in the thickness of the core layer results in a drop in non-dimensional frequencies, hence enhancing the system's stability. Furthermore, it is disclosed that the frequency of the structure can be regulated by altering the wall thickness of the auxetic lattice. An increase in the thickness-to-inclined length ratio leads to a decrease in the plate's frequency. Due to the enhancement that can be achieved in both longitudinal and torsional rigidity of the plate.

Figure 3.10 has been presented to compare the effects of several kinds of the plate for two distinct degrees of velocity feedback gain. The addition of a magnetostrictive layer to the plate increases the dimensionless natural frequency, as evidenced by these data. The pure magnetostrictive plate exhibits the highest natural frequency among the three types.

3.11 Conclusion

This work sought to examine the active control behavior of plates with an Auxetic core and coated with magnetostrictive materials, employing the First-order Shear Deformation Theory (FSDT). This action was executed to improve understanding of the phenomenon. The governing equations

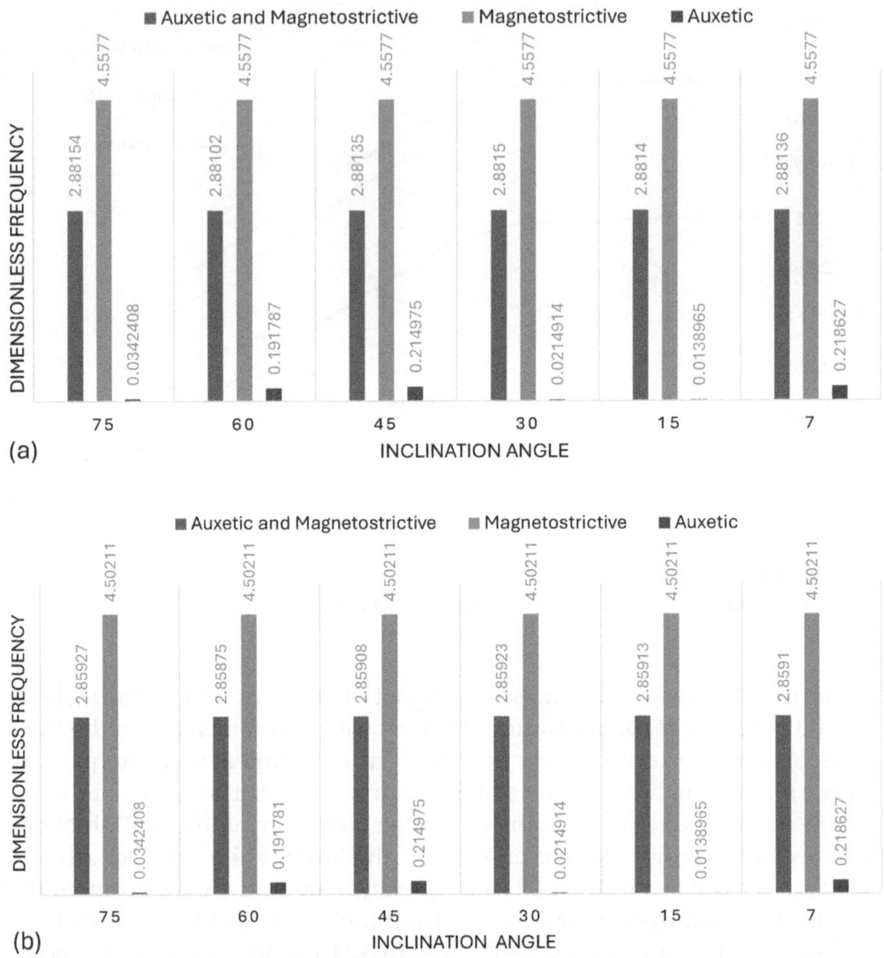

FIGURE 3.10
Comparison of dimensionless natural frequencies for three distinct plate types (a) velocity feedback gain = 5 x 10^5 (b) velocity feedback gain = 9 x 10^5.

were derived using Hamilton's principle, and Navier's analytical solution was subsequently applied to resolve these equations. The researchers discerned numerous significant things from the findings of the current investigation.

- Of the three plate types examined in this work, the magnetostrictive plate demonstrates the greatest dimensionless natural frequency values.

- The dimensionless natural frequencies diminish with an increase in the inclination angle.
- The dimensionless natural frequency increases with the elongation of the rib.
- Magnetostrictive material influences the behavior of the plate.

Appendix A

$$K_{11} = -A_{11}\left(o^2\right)$$

$$K_{12} = -B_{11}\left(o^2\right) + \alpha S_{11}\left(o^2\right)$$

$$K_{13} = \alpha S_{11}\left(o^3\right)$$

$$K_{21} = -B_{11}\left(o^2\right) + \alpha S_{11}\left(o^2\right)$$

$$K_{22} = -D_{11}\left(o^2\right) + 2\alpha M_{11}\left(o^2\right) - \alpha^2 L_{11}\left(o^2\right) - A_{55} + 2\beta D_{55} - \beta^2 M_{55}$$

$$K_{23} = \alpha M_{11}\left(o^3\right) - \alpha^2 L_{11}\left(o^3\right) + 2\beta D_{55}(o) - \beta^2 M_{55}(o) - A_{55}(o)$$

$$K_{31} = \alpha S_{11}\left(o^3\right)$$

$$K_{32} = \alpha M_{11}\left(o^3\right) - \alpha^2 L_{11}\left(o^3\right) + 2\beta D_{55}(o) - \beta^2 M_{55}(o) - A_{55}(o)$$

$$K_{33} = -\alpha^2 L_{11}\left(o^4\right) - \beta^2 M_{55}\left(o^2\right) + 2\beta D_{55}\left(o^2\right) - A_{55}\left(o^2\right) - K_w - K_g\left(o^2\right) - K_w\left(\mu^2 o^2\right) - K_g\left(\mu^2 o^4\right)$$

$$M_{11} = I_0\left(1 + \mu^2 o^2\right)$$

$$M_{12} = -\left(I_1 - I_3\alpha + \mu^2 I_1 o^2 + I_3\alpha\,\mu^2 o^2\right)$$

$$M_{13} = -\left(-I_3\alpha(o) - \mu^2 I_3\alpha\left(o^3\right)\right)$$

$$M_{21} = -\left(I_1 - I_3\alpha + \mu^2 I_1 o^2 + I_3\alpha\,\mu^2 o^2\right)$$

$$M_{22} = -\left(I_2 - 2I_4\alpha + I_6\alpha^2 + I_2\mu^2 o^2 - 2I_4\alpha o^2 + I_6\mu^2\alpha^2 o^2\right)$$

$$M_{23} = -\left(I_6\alpha^2 o - I_4\alpha(o) + I_6\alpha^2\mu^2 o^3 - I_4\alpha\mu^2 o^3\right)$$

$$M_{31} = -\left(-I_3\alpha(o) - \mu^2 I_3\alpha\left(o^3\right)\right)$$

$$M_{32}=-\left(I_6\alpha^2 o-I_4\alpha(o)+I_6\alpha^2\mu^2 o^3-I_4\alpha\mu^2 o^3\right)$$

$$M_{33}=-\left(I_0+I_6 o^2\alpha^2+I_0\mu^2 o^2+I_6\mu^2 o^4\alpha^2\right)$$

$$A_{11}=\int_{-h_c/2}^{h_c/2}C_{11}^c dz+\int_{-\frac{(hc+h_f)}{2}}^{\frac{-hc}{2}}C_{11}^f dz+\int_{\frac{h_c}{2}}^{\frac{(h_c+h_f)}{2}}C_{11}^f dz$$

$$A_{12}=\int_{-h_c/2}^{h_c/2}C_{12}^c dz+\int_{-\frac{(hc+h_f)}{2}}^{\frac{-hc}{2}}C_{12}^f dz+\int_{\frac{h_c}{2}}^{\frac{(h_c+h_f)}{2}}C_{12}^f dz$$

$$A_{22}=\int_{-h_c/2}^{h_c/2}C_{22}^c dz+\int_{-\frac{(hc+h_f)}{2}}^{\frac{-hc}{2}}C_{22}^f dz+\int_{\frac{h_c}{2}}^{\frac{(h_c+h_f)}{2}}C_{22}^f dz$$

$$A_{44}=\int_{-h_c/2}^{h_c/2}C_{44}^c dz+\int_{-\frac{(hc+h_f)}{2}}^{\frac{-hc}{2}}C_{44}^f dz+\int_{\frac{h_c}{2}}^{\frac{(h_c+h_f)}{2}}C_{44}^f dz$$

$$A_{55}=\int_{-h_c/2}^{h_c/2}C_{55}^c dz+\int_{-\frac{(hc+h_f)}{2}}^{\frac{-hc}{2}}C_{55}^f dz+\int_{\frac{h_c}{2}}^{\frac{(h_c+h_f)}{2}}C_{55}^f dz$$

$$A_{66}=\int_{-h_c/2}^{h_c/2}C_{66}^c dz+\int_{-\frac{(hc+h_f)}{2}}^{\frac{-hc}{2}}C_{66}^f dz+\int_{\frac{h_c}{2}}^{\frac{(h_c+h_f)}{2}}C_{66}^f dz$$

$$B_{11}=\int_{-h_c/2}^{h_c/2}zC_{11}^c dz+\int_{-\frac{(hc+h_f)}{2}}^{\frac{-hc}{2}}zC_{11}^f dz+\int_{\frac{h_c}{2}}^{\frac{(h_c+h_f)}{2}}zC_{11}^f dz$$

$$B_{12}=\int_{-h_c/2}^{h_c/2}zC_{12}^c dz+\int_{-\frac{(hc+h_f)}{2}}^{\frac{-hc}{2}}zC_{12}^f dz+\int_{\frac{h_c}{2}}^{\frac{(h_c+h_f)}{2}}zC_{12}^f dz$$

$$B_{22}=\int_{-h_c/2}^{h_c/2}zC_{22}^c dz+\int_{-\frac{(hc+h_f)}{2}}^{\frac{-hc}{2}}zC_{22}^f dz+\int_{\frac{h_c}{2}}^{\frac{(h_c+h_f)}{2}}zC_{22}^f dz$$

$$B_{44} = \int_{-h_c/2}^{h_c/2} zC^{c}_{44}dz + \int_{-\frac{(hc+h_f)}{2}}^{\frac{-hc}{2}} zC^{f}_{44}dz + \int_{\frac{h_c}{2}}^{\frac{(h_c+h_f)}{2}} zC^{f}_{44}dz$$

$$B_{55} = \int_{-h_c/2}^{h_c/2} zC^{c}_{55}dz + \int_{-\frac{(hc+h_f)}{2}}^{\frac{-hc}{2}} zC^{f}_{22}dz + \int_{\frac{h_c}{2}}^{\frac{(h_c+h_f)}{2}} zC^{f}_{55}dz$$

$$B_{66} = \int_{-h_c/2}^{h_c/2} zC^{c}_{66}dz + \int_{-\frac{(hc+h_f)}{2}}^{\frac{-hc}{2}} zC^{f}_{66}dz + \int_{\frac{h_c}{2}}^{\frac{(h_c+h_f)}{2}} zC^{f}_{66}dz$$

$$D_{11} = \int_{-h_c/2}^{h_c/2} z^2C^{c}_{11}dz + \int_{-\frac{(hc+h_f)}{2}}^{\frac{-hc}{2}} z^2C^{f}_{11}dz + \int_{\frac{h_c}{2}}^{\frac{(h_c+h_f)}{2}} z^2C^{f}_{11}dz$$

$$D_{12} = \int_{-h_c/2}^{h_c/2} z^2C^{c}_{12}dz + \int_{-\frac{(hc+h_f)}{2}}^{\frac{-hc}{2}} z^2C^{f}_{12}dz + \int_{\frac{h_c}{2}}^{\frac{(h_c+h_f)}{2}} z^2C^{f}_{12}dz$$

$$D_{22} = \int_{-h_c/2}^{h_c/2} z^2C^{c}_{22}dz + \int_{-\frac{(hc+h_f)}{2}}^{\frac{-hc}{2}} z^2C^{f}_{22}dz + \int_{\frac{h_c}{2}}^{\frac{(h_c+h_f)}{2}} z^2C^{f}_{22}dz$$

$$D_{44} = \int_{-h_c/2}^{h_c/2} z^2C^{c}_{44}dz + \int_{-\frac{(hc+h_f)}{2}}^{\frac{-hc}{2}} z^2C^{f}_{44}dz + \int_{\frac{h_c}{2}}^{\frac{(h_c+h_f)}{2}} z^2C^{f}_{44}dz$$

$$D_{55} = \int_{-h_c/2}^{h_c/2} z^2C^{c}_{55}dz + \int_{-\frac{(hc+h_f)}{2}}^{\frac{-hc}{2}} z^2C^{f}_{55}dz + \int_{\frac{h_c}{2}}^{\frac{(h_c+h_f)}{2}} z^2C^{f}_{55}dz$$

$$D_{66} = \int_{-h_c/2}^{h_c/2} z^2C^{c}_{66}dz + \int_{-\frac{(hc+h_f)}{2}}^{\frac{-hc}{2}} z^2C^{f}_{66}dz + \int_{\frac{h_c}{2}}^{\frac{(h_c+h_f)}{2}} z^2C^{f}_{66}dz$$

$$S_{11} = \int_{-h_c/2}^{h_c/2} z^3C^{c}_{11}dz + \int_{-\frac{(hc+h_f)}{2}}^{\frac{-hc}{2}} z^3C^{f}_{11}dz + \int_{\frac{h_c}{2}}^{\frac{(h_c+h_f)}{2}} z^3C^{f}_{11}dz$$

$$S_{12} = \int_{-h_c/2}^{h_c/2} z^3 C_{12}^{c} dz + \int_{-\frac{(hc+h_f)}{2}}^{\frac{-hc}{2}} z^3 C_{12}^{f} dz + \int_{\frac{h_c}{2}}^{\frac{(h_c+h_f)}{2}} z^3 C_{12}^{f} dz$$

$$S_{22} = \int_{-h_c/2}^{h_c/2} z^3 C_{22}^{c} dz + \int_{-\frac{(hc+h_f)}{2}}^{\frac{-hc}{2}} z^3 C_{22}^{f} dz + \int_{\frac{h_c}{2}}^{\frac{(h_c+h_f)}{2}} z^3 C_{22}^{f} dz$$

$$S_{44} = \int_{-h_c/2}^{h_c/2} z^3 C_{44}^{c} dz + \int_{-\frac{(hc+h_f)}{2}}^{\frac{-hc}{2}} z^3 C_{44}^{f} dz + \int_{\frac{h_c}{2}}^{\frac{(h_c+h_f)}{2}} z^3 C_{44}^{f} dz$$

$$S_{55} = \int_{-h_c/2}^{h_c/2} z^3 C_{55}^{c} dz + \int_{-\frac{(hc+h_f)}{2}}^{\frac{-hc}{2}} z^3 C_{55}^{f} dz + \int_{\frac{h_c}{2}}^{\frac{(h_c+h_f)}{2}} z^3 C_{55}^{f} dz$$

$$S_{66} = \int_{-h_c/2}^{h_c/2} z^3 C_{66}^{c} dz + \int_{-\frac{(hc+h_f)}{2}}^{\frac{-hc}{2}} z^3 C_{66}^{f} dz + \int_{\frac{h_c}{2}}^{\frac{(h_c+h_f)}{2}} z^3 C_{66}^{f} dz$$

$$M_{11} = \int_{-h_c/2}^{h_c/2} z^4 C_{11}^{c} dz + \int_{-\frac{(hc+h_f)}{2}}^{\frac{-hc}{2}} z^4 C_{11}^{f} dz + \int_{\frac{h_c}{2}}^{\frac{(h_c+h_f)}{2}} z^4 C_{11}^{f} dz$$

$$M_{12} = \int_{-h_c/2}^{h_c/2} z^4 C_{12}^{c} dz + \int_{-\frac{(hc+h_f)}{2}}^{\frac{-hc}{2}} z^4 C_{12}^{f} dz + \int_{\frac{h_c}{2}}^{\frac{(h_c+h_f)}{2}} z^4 C_{12}^{f} dz$$

$$M_{22} = \int_{-h_c/2}^{h_c/2} z^4 C_{22}^{c} dz + \int_{-\frac{(hc+h_f)}{2}}^{\frac{-hc}{2}} z^4 C_{22}^{f} dz + \int_{\frac{h_c}{2}}^{\frac{(h_c+h_f)}{2}} z^4 C_{22}^{f} dz$$

$$M_{44} = \int_{-h_c/2}^{h_c/2} z^4 C_{44}^{c} dz + \int_{-\frac{(hc+h_f)}{2}}^{\frac{-hc}{2}} z^4 C_{44}^{f} dz + \int_{\frac{h_c}{2}}^{\frac{(h_c+h_f)}{2}} z^4 C_{44}^{f} dz$$

$$M_{55} = \int_{-h_c/2}^{h_c/2} z^4 C_{55}^{c} dz + \int_{-\frac{(hc+h_f)}{2}}^{\frac{-hc}{2}} z^4 C_{22}^{f} dz + \int_{\frac{h_c}{2}}^{\frac{(h_c+h_f)}{2}} z^4 C_{55}^{f} dz$$

$$M_{66}=\int_{-h_c/2}^{h_c/2} z^4 C_{66}^c dz+\int_{-\frac{(hc+h_f)}{2}}^{\frac{-hc}{2}} z^4 C_{66}^f dz+\int_{\frac{h_c}{2}}^{\frac{(h_c+h_f)}{2}} z^4 C_{66}^f dz$$

$$L_{11}=\int_{-h_c/2}^{h_c/2} z^6 C_{11}^c dz+\int_{-\frac{(hc+h_f)}{2}}^{\frac{-hc}{2}} z^6 C_{11}^f dz+\int_{\frac{h_c}{2}}^{\frac{(h_c+h_f)}{2}} z^6 C_{11}^f dz$$

$$L_{12}=\int_{-h_c/2}^{h_c/2} z^6 C_{12}^c dz+\int_{-\frac{(hc+h_f)}{2}}^{\frac{-hc}{2}} z^6 C_{12}^f dz+\int_{\frac{h_c}{2}}^{\frac{(h_c+h_f)}{2}} z^6 C_{12}^f dz$$

$$L_{22}=\int_{-h_c/2}^{h_c/2} z^6 C_{22}^c dz+\int_{-\frac{(hc+h_f)}{2}}^{\frac{-hc}{2}} z^6 C_{22}^f dz+\int_{\frac{h_c}{2}}^{\frac{(h_c+h_f)}{2}} z^6 C_{22}^f dz$$

$$L_{44}=\int_{-h_c/2}^{h_c/2} z^6 C_{44}^c dz+\int_{-\frac{(hc+h_f)}{2}}^{\frac{-hc}{2}} z^6 C_{44}^f dz+\int_{\frac{h_c}{2}}^{\frac{(h_c+h_f)}{2}} z^6 C_{44}^f dz$$

$$L_{55}=\int_{-h_c/2}^{h_c/2} z^6 C_{55}^c dz+\int_{-\frac{(hc+h_f)}{2}}^{\frac{-hc}{2}} z^6 C_{55}^f dz+\int_{\frac{h_c}{2}}^{\frac{(h_c+h_f)}{2}} z^6 C_{55}^f dz$$

$$L_{66}=\int_{-h_c/2}^{h_c/2} z^6 C_{66}^c dz+\int_{-\frac{(hc+h_f)}{2}}^{\frac{-hc}{2}} z^6 C_{66}^f dz+\int_{\frac{h_c}{2}}^{\frac{(h_c+h_f)}{2}} z^6 C_{66}^f dz$$

Appendix B

$$K_{11} = -A_{11}\left(\alpha^2\right) - A_{66}\left(\beta^2\right) \qquad K_{12} = -A_{12}\left(\alpha\beta\right) - A_{66}\left(\alpha\beta\right)$$

$$K_{13} = 0 \qquad K_{14} = -B_{11}\left(\alpha^2\right) - B_{66}\left(\beta^2\right)$$

$$K_{15} = -B_{12}\left(\alpha\beta\right) - B_{66}\left(\alpha\beta\right)$$

$$K_{21} = -A_{21}\left(\alpha\beta\right) - A_{66}\left(\alpha\beta\right) \qquad K_{22} = -A_{22}\left(\alpha^2\right) - A_{66}\left(\beta^2\right)$$

$$K_{23} = 0 \qquad K_{24} = -B_{21}\left(\alpha\beta\right) - B_{66}\left(\alpha\beta\right)$$

$$K_{25} = -B_{22}\left(\alpha^2\right) - B_{66}\left(\beta^2\right)$$

$$K_{31} = 0 \qquad K_{32} = 0$$

$$K_{33} = -A_{55}\left(\alpha^2\right) - A_{44}\left(\beta^2\right) - K_w - K_g\left(\alpha^2\right) - K_g\left(\beta^2\right) - K_w\left(\mu^2\alpha^2\right) - K_w\left(\mu^2\beta^2\right) - K_g\left(\mu^2\alpha^4\right) - K_g\left(\mu^2\beta^4\right) - 2K_g\left(\mu^2\alpha^2\beta^2\right)$$

$$K_{34} = -A_{55}\left(\alpha\right) \qquad K_{35} = -A_{44}\left(\beta\right)$$

$$K_{41} = -B_{11}\left(\alpha^2\right) - B_{66}\left(\beta^2\right) \qquad K_{42} = -B_{21}\left(\alpha\beta\right) - B_{66}\left(\alpha\beta\right)$$

$$K_{43} = -A_{55}\left(\alpha\right) \qquad K_{44} = -D_{11}\left(\alpha^2\right) - D_{66}\left(\beta^2\right) - A_{55}$$

$$K_{45} = -D_{12}\left(\alpha\beta\right) - D_{66}\left(\alpha\beta\right)$$

$$K_{51} = -B_{12}\left(\alpha\beta\right) - B_{66}\left(\alpha\beta\right) \qquad K_{52} = -B_{22}\left(\alpha^2\right) - B_{66}\left(\beta^2\right)$$

$$K_{53} = -A_{44}\left(\beta\right) \qquad K_{54} = -D_{12}\left(\alpha\beta\right) - D_{66}\left(\alpha\beta\right)$$

$$K_{55} = -D_{22}\left(\beta^2\right) - D_{66}\left(\alpha^2\right) - A_{44}$$

$$M_{11} = I_0\left(1 + \mu^2\alpha^2 + \mu^2\beta^2\right) \qquad M_{14} = I_1\left(1 + \mu^2\alpha^2 + \mu^2\beta^2\right)$$

$$M_{22} = I_0\left(1+\mu^2\alpha^2+\mu^2\beta^2\right) \qquad M_{25} = I_1\left(1+\mu^2\alpha^2+\mu^2\beta^2\right)$$

$$M_{33} = I_0\left(1+\mu^2\alpha^2+\mu^2\beta^2\right) \qquad M_{41} = I_1\left(1+\mu^2\alpha^2+\mu^2\beta^2\right)$$

$$M_{44} = I_2\left(1+\mu^2\alpha^2+\mu^2\beta^2\right) \qquad M_{52} = I_1\left(1+\mu^2\alpha^2+\mu^2\beta^2\right)$$

$$M_{55} = I_2\left(1+\mu^2\alpha^2+\mu^2\beta^2\right)$$

$$C_{11} = C_{12} = C_{14} = C_{15} = 0$$

$$C_{13} = \int_{-\frac{(h_a+h_m)}{2}}^{\frac{-h_a}{2}} e_{31}(\alpha)K_cC(t)dz + \int_{\frac{h_a}{2}}^{\frac{(h_a+h_m)}{2}} e_{31}(\alpha)K_cC(t)dz$$

$$C_{21} = C_{22} = C_{24} = C_{25} = 0$$

$$C_{23} = \int_{-\frac{(h_a+h_m)}{2}}^{\frac{-h_a}{2}} e_{32}(\beta)K_cC(t)dz + \int_{\frac{h_a}{2}}^{\frac{(h_a+h_m)}{2}} e_{32}(\beta)K_cC(t)dz$$

$$C_{31} = C_{13} \qquad C_{32} = C_{23}$$

$$C_{33} = 0$$

$$C_{34} = \int_{-\frac{(h_a+h_m)}{2}}^{\frac{-h_a}{2}} ze_{31}(\alpha)K_cC(t)dz + \int_{\frac{h_a}{2}}^{\frac{(h_a+h_m)}{2}} ze_{31}(\alpha)K_cC(t)dz$$

$$C_{35} = \int_{-\frac{h_c}{2}}^{\frac{h_c}{2}} ze_{32}(\beta)K_cC(t)dz$$

$$C_{41} = C_{42} = C_{44} = C_{45} = 0 \qquad C_{43} = C_{34}$$

$$C_{51} = C_{52} = C_{54} = C_{55} = 0$$

$$C_{53} = \int_{-\frac{(h_a+h_m)}{2}}^{\frac{-h_a}{2}} ze_{32}(\beta)K_cC(t)dz + \int_{\frac{h_a}{2}}^{\frac{(h_a+h_m)}{2}} ze_{32}(\beta)K_cC(t)dz$$

$$A_{11} = \int_{-h_a/2}^{h_a/2} C_{11}^a dz + \int_{-\frac{(h_a+h_m)}{2}}^{\frac{-h_a}{2}} C_{11}^m dz + \int_{\frac{h_a}{2}}^{\frac{(h_a+h_m)}{2}} C_{11}^m dz$$

$$A_{12} = \int_{-h_a/2}^{h_a/2} C_{12}^a dz + \int_{-\frac{(h_a+h_m)}{2}}^{\frac{-h_a}{2}} C_{12}^m dz + \int_{\frac{h_c}{2}}^{\frac{(h_a+h_m)}{2}} C_{12}^m dz$$

$$A_{22} = \int_{-h_a/2}^{h_a/2} C_{22}^a dz + \int_{-\frac{(h_a+h_m)}{2}}^{\frac{-h_a}{2}} C_{22}^m dz + \int_{\frac{h_a}{2}}^{\frac{(h_a+h_m)}{2}} C_{22}^m dz$$

$$A_{44} = \int_{-h_a/2}^{h_a/2} C_{44}^a dz + \int_{-\frac{(h_a+h_m)}{2}}^{\frac{-h_a}{2}} C_{44}^m dz + \int_{\frac{h_a}{2}}^{\frac{(h_a+h_m)}{2}} C_{44}^m dz$$

$$A_{55} = \int_{-h_a/2}^{h_a/2} C_{55}^a dz + \int_{-\frac{(h_a+h_m)}{2}}^{\frac{-h_a}{2}} C_{55}^m dz + \int_{\frac{h_a}{2}}^{\frac{(h_c+h_m)}{2}} C_{55}^m dz$$

$$A_{66} = \int_{-h_a/2}^{h_a/2} C_{66}^a dz + \int_{-\frac{(h_a+h_m)}{2}}^{\frac{-h_a}{2}} C_{66}^m dz + \int_{\frac{h_a}{2}}^{\frac{(h_a+h_m)}{2}} C_{66}^m dz$$

$$B_{11} = \int_{-h_a/2}^{h_a/2} zC_{11}^a dz + \int_{-\frac{(hc+h_m)}{2}}^{\frac{-h_a}{2}} zC_{11}^m dz + \int_{\frac{h_c}{2}}^{\frac{(h_a+h_m)}{2}} zC_{11}^m dz$$

$$B_{12} = \int_{-h_a/2}^{h_a/2} zC_{12}^a dz + \int_{-\frac{(h_a+h_m)}{2}}^{\frac{-h_a}{2}} zC_{12}^m dz + \int_{\frac{h_a}{2}}^{\frac{(h_a+h_m)}{2}} zC_{12}^m dz$$

$$B_{22} = \int_{-h_a/2}^{h_a/2} zC_{22}^a dz + \int_{-\frac{(h_a+h_m)}{2}}^{\frac{-h_a}{2}} zC_{22}^m dz + \int_{\frac{h_a}{2}}^{\frac{(h_a+h_m)}{2}} zC_{22}^m dz$$

$$B_{44} = \int_{-h_a/2}^{h_a/2} zC_{44}^a dz + \int_{-\frac{(h_a+h_m)}{2}}^{\frac{-h_a}{2}} zC_{44}^m dz + \int_{\frac{h_a}{2}}^{\frac{(h_a+h_m)}{2}} zC_{44}^m dz$$

$$B_{55} = \int_{-h_a/2}^{h_a/2} zC_{55}^a dz + \int_{-\frac{(h_a+h_m)}{2}}^{\frac{-h_a}{2}} zC_{22}^m dz + \int_{\frac{h_a}{2}}^{\frac{(h_a+h_m)}{2}} zC_{55}^m dz$$

$$B_{66}=\int_{-h_a/2}^{h_a/2} zC_{66}^{a}dz+\int_{-\frac{(h_a+h_m)}{2}}^{\frac{-h_a}{2}} zC_{66}^{m}dz+\int_{\frac{h_a}{2}}^{\frac{(h_c+h_m)}{2}} zC_{66}^{m}dz$$

$$D_{11}=\int_{-h_a/2}^{h_a/2} z^2C_{11}^{a}dz+\int_{-\frac{(h_a+h_f)}{2}}^{\frac{-h_a}{2}} z^2C_{11}^{m}dz+\int_{\frac{h_a}{2}}^{\frac{(h_a+h_f)}{2}} z^2C_{11}^{m}dz$$

$$D_{12}=\int_{-h_a/2}^{h_a/2} z^2C_{12}^{a}dz+\int_{-\frac{(h_a+h_m)}{2}}^{\frac{-h_a}{2}} z^2C_{12}^{m}dz+\int_{\frac{h_a}{2}}^{\frac{(h_a+h_m)}{2}} z^2C_{12}^{m}dz$$

$$D_{22}=\int_{-h_a/2}^{h_a/2} z^2C_{22}^{a}dz+\int_{-\frac{(h_a+h_m)}{2}}^{\frac{-h_a}{2}} z^2C_{22}^{m}dz+\int_{\frac{h_a}{2}}^{\frac{(h_a+h_m)}{2}} z^2C_{22}^{m}dz$$

$$D_{44}=\int_{-h_a/2}^{h_a/2} z^2C_{44}^{a}dz+\int_{-\frac{(h_a+h_m)}{2}}^{\frac{-h_a}{2}} z^2C_{44}^{m}dz+\int_{\frac{h_a}{2}}^{\frac{(h_a+h_m)}{2}} z^2C_{44}^{m}dz$$

$$D_{55}=\int_{-h_a/2}^{h_a/2} z^2C_{55}^{a}dz+\int_{-\frac{(h_a+h_m)}{2}}^{\frac{-h_a}{2}} z^2C_{55}^{m}dz+\int_{\frac{h_a}{2}}^{\frac{(h_a+h_m)}{2}} z^2C_{55}^{m}dz$$

$$D_{66}=\int_{-h_a/2}^{h_a/2} z^2C_{66}^{a}dz+\int_{-\frac{(h_a+h_m)}{2}}^{\frac{-h_a}{2}} z^2C_{66}^{m}dz+\int_{\frac{h_a}{2}}^{\frac{(h_a+h_m)}{2}} z^2C_{66}^{m}dz$$

References

1. Ebrahimi, F., and M.F. Ahari, Dynamic analysis of meta-material plates with magnetostrictive face sheets. *International Journal of Structural Stability and Dynamics*, 2024. 24(16): p. 2450174.
2. Ebrahimi, F., and M.F. Ahari, Vibration control of graphene-origami-enabled meta-material beams with magnetostrictive coatings under temperature gradient. *Acta Mechanica*, 2024. 235(4): p. 2193–2209.
3. Ebrahimi, F., and M.F. Ahari, On the buckling of meta-graphene-origami-enabled magnetostrictive nanoplates under temperature gradient. *Acta Mechanica*, 2024: p. 1–18.
4. Ebrahimi, F., and M.F. Ahari, Meta-material beams with magnetostrictive coatings: vibrational characteristics. *Acta Mechanica*, 2024: p. 1–18.

5. Ebrahimi, F., and M.F. Ahari, Thermomechanical active vibration control of auxetic plates with magnetostrictive layers. *Journal of the Brazilian Society of Mechanical Sciences and Engineering*, 2024. 46(1): p. 19.
6. Ghadiri, M., M.F. Ahari, and M. Marvi, Auxetic metamaterial pre-twisted helical nanobeams: vibrational characteristics. *Journal of the Brazilian Society of Mechanical Sciences and Engineering*, 2024. 46(7): p. 436.
7. Ebrahimi, F., and M.F. Ahari, *Mechanics of Magnetostrictive Materials and Structures*. 2023: CRC Press.
8. Rao, S.S., *Vibration of Continuous Systems*. 2019: John Wiley & Sons.
9. Ebrahimi, F., and M.F. Ahari, Magnetostriction-assisted active control of the multi-layered nanoplates: effect of the porous functionally graded facesheets on the system's behavior. *Engineering with Computers*, 2023: p. 1–15.
10. Ebrahimi, F., M.F. Ahari, and A. Dabbagh, Stability analysis of a sandwich composite magnetostrictive nanoplate coupled with FG porous facesheets. *Acta Mechanica*, 2024: p. 1–23.
11. Ahari, M.F., and M. Ghadiri, Resonator vibration of a magneto-electro-elastic nano-plate integrated with FGM layer subjected to the nano mass-Spring-damper system and a moving load. *Waves in Random and Complex Media*, 2022: p. 1–39.
12. Ebrahimi, F., and M.F. Ahari, Dynamic analysis of sandwich magnetostrictive nanoplates with a mass – spring – damper stimulator. *International Journal of Structural Stability and Dynamics*, 2024. 24(09): p. 2450136.

4

Buckling of Active Material Nanoplates

4.1 Background

Buckling is a type of mechanical instability that causes an instantaneous out-of-plane deformation in a thin plate or shell subjected to compressive loading. At the nanoscale, these traditional continuum theories fail to capture size-dependent behavior, and thus the use of nonlocal elasticity, surface elasticity, and strain gradient theories is needed to model the behavior accurately. These effects are particularly significant in nanoplates, where material behavior such as stiffness, critical buckling load, and post-buckling response is strongly influenced by scale-dependent behavior.

Buckling of nanoplates has gained widespread interest in recent years due to its central significance in the high-performance design of nanoscale devices and structures. Active material nanoplates, with their reversible changes in mechanical properties under external stimuli such as electric, magnetic, or thermal fields, represent particular challenges and opportunities in engineering applications.

4.2 Theoretical Framework and Formulation

We will analyze a rectangular plate defined by its dimensions: length (a), width (b), and total thickness (h) in the x, y, and z axes, respectively. This plate consists of nano-faces with magnetostrictive characteristics and a core layer demonstrating auxetic behavior. The current study utilizes the Cartesian coordinate system, illustrated in Figure 4.1, with the origin located at one corner of the plate. The methodology developed by Winkler and Pasternak has been employed in the analysis of foundation effect phenomena.

DOI: 10.1201/9781003540731-4

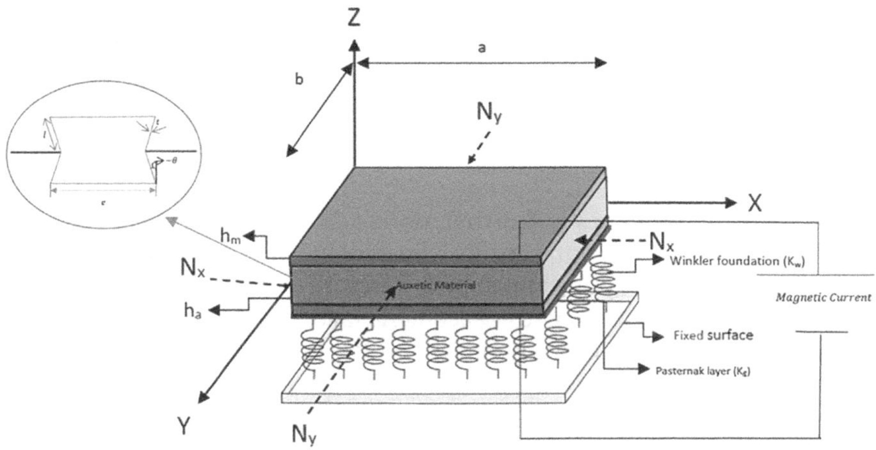

FIGURE 4.1
Arrangement of sandwich composite.

4.2.1 Auxetic Materials

The auxetic core layer consists of cells arranged in a honeycomb configuration. The mechanical properties of the honeycomb auxetic core layer, including Young's modulus, shear modulus, negative Poisson's ratio, and thermal expansion coefficients, are thought to be affected by the geometrical features of the individual unit cell.

4.2.1.1 Auxetic Core

The possible ramifications regarding the efficient mechanical properties of the auxetic core are as Eq. (3–4) through Eq. (3–8).

4.2.2 Geometric Configuration

This study utilizes the advanced sinusoidal shear deformation plate theory initially proposed by Touratier [1]. According to the concepts of higher-order sinusoidal shear deformation plate theory, the normal direction to the midline, referred to as the neutral surface, retains its perpendicular orientation both before and following deformation.

4.2.2.1 Displacement field

The displacement field can be articulated according to the principles of higher-order sinusoidal shear deformation plate theory [2]:

$$U_x\left(x,y,z,t\right)=u_0\left(x,y,t\right)-z\frac{\partial w_0\left(x,y,t\right)}{\partial x}+f\left(z\right)\phi_x \tag{4-1}$$

$$V_y(x,y,z,t) = v_0(x,y,t) - z\frac{\partial w_0(x,y,t)}{\partial y} + f(z)\phi_y(x,y,t) \tag{4-2}$$

$$W_Z(x,y,z,t) = w_0(x,y,t) \tag{4-3}$$

It is essential to acknowledge that, unlike the first-order shear deformation plate theory, the higher-order shear deformation theory eliminates the necessity of incorporating a shear correction factor. Furthermore, the function $f(z)$ denotes the displacement field in the direction orthogonal to the surface, specifically along the thickness axis, and is formally expressed as follows: $f(z) = \frac{H}{\pi}\sin\left(\frac{\pi z}{H}\right), g(z) = \frac{df(z)}{dz} = \cos\left(\frac{\pi z}{H}\right)$

4.2.3 Eringen's Nonlocal Theory

According to Eringen's nonlocal theory, applicable in micro and nanoscale contexts, the stress at a certain reference point (x) inside an elastic continuum is impacted not only by the strain at point x but also by the strains at all neighboring points. This relationship can be articulated as Eq. (3–9) [3].

4.2.4 Stress-Strain Relationship for Magnetostrictive Materials

The distinctive characteristics of magnetostrictive materials have garnered significant attention in engineering because of their capacity to transform energy between various forms (magnetic, electric, or mechanical). Eqs. (3–43) to (3–47) represents the stress, strain, and magnetic susceptibility of the magnetostrictive layer [2–4]:

The magnetic field (H_{zz}) can be calculated using the Eq. (3–13):

The non-zero strains of the magnetostrictive layer may be characterized as [2]:

$$\begin{Bmatrix} \varepsilon_{xx}^m \\ \varepsilon_{yy}^m \\ \varepsilon_{yz}^m \\ \varepsilon_{xz}^m \\ \varepsilon_{xy}^m \end{Bmatrix} = \begin{Bmatrix} \varepsilon_{xx}^m \\ \varepsilon_{yy}^m \\ \varepsilon_{yz}^m \\ \varepsilon_{xz}^m \\ \varepsilon_{xy}^m \end{Bmatrix} + z\begin{Bmatrix} \kappa_{xx}^m \\ \kappa_{yy}^m \\ \kappa_{yz}^m \\ \kappa_{xz}^m \\ \kappa_{xy}^m \end{Bmatrix} + f(z)\begin{Bmatrix} \chi_{xx}^m \\ \chi_{yy}^m \\ \chi_{yz}^m \\ \chi_{xz}^m \\ \chi_{xy}^m \end{Bmatrix} + f(z)'\begin{Bmatrix} X_{xx}^m \\ X_{yy}^m \\ X_{yz}^m \\ X_{xz}^m \\ X_{xy}^m \end{Bmatrix} \tag{4-4}$$

in which,

$$\begin{Bmatrix} \varepsilon_{xx}^{m} \\ \varepsilon_{yy}^{m} \\ \varepsilon_{yz}^{m} \\ \varepsilon_{xz}^{m} \\ \varepsilon_{xy}^{m} \end{Bmatrix} = \begin{Bmatrix} \dfrac{\partial u_0}{\partial x} \\ \dfrac{\partial v_0}{\partial y} \\ 0 \\ 0 \\ \left(\dfrac{\partial u_0}{\partial y} + \dfrac{\partial v_0}{\partial x} \right) \end{Bmatrix}, \begin{Bmatrix} \kappa_{xx}^{m} \\ \kappa_{yy}^{m} \\ \kappa_{yz}^{m} \\ \kappa_{xz}^{m} \\ \kappa_{xy}^{m} \end{Bmatrix} = \begin{Bmatrix} -\dfrac{\partial^2 w_0}{\partial x^2} \\ -\dfrac{\partial^2 w_0}{\partial y^2} \\ 0 \\ 0 \\ -\left(\dfrac{\partial^2 w_0}{\partial x^2} + \dfrac{\partial^2 w_0}{\partial y^2} \right) \end{Bmatrix}, \begin{Bmatrix} \chi_{xx}^{m} \\ \chi_{yy}^{m} \\ \chi_{yz}^{m} \\ \chi_{xz}^{m} \\ \chi_{xy}^{m} \end{Bmatrix} = \begin{Bmatrix} \dfrac{\partial \phi_x}{\partial x} \\ \dfrac{\partial \phi_y}{\partial y} \\ 0 \\ 0 \\ \left(\dfrac{\partial \phi_x}{\partial y} + \dfrac{\partial \phi_y}{\partial x} \right) \end{Bmatrix}$$

(4-5)

4.2.5 Stress-Strain Relationship of Auxetic Materials

The non-zero stresses of the auxetic layer are defined as follows:

$$\begin{Bmatrix} \varepsilon_{xx}^{a} \\ \varepsilon_{yy}^{a} \\ \varepsilon_{yz}^{a} \\ \varepsilon_{xz}^{a} \\ \varepsilon_{xy}^{a} \end{Bmatrix} = \begin{Bmatrix} \varepsilon_{xx}^{a} \\ \varepsilon_{yy}^{a} \\ \varepsilon_{yz}^{a} \\ \varepsilon_{xz}^{a} \\ \varepsilon_{xy}^{a} \end{Bmatrix} + z \begin{Bmatrix} \kappa_{xx}^{a} \\ \kappa_{yy}^{a} \\ \kappa_{yz}^{a} \\ \kappa_{xz}^{a} \\ \kappa_{xy}^{a} \end{Bmatrix} \tag{4-6}$$

in which,

$$\begin{Bmatrix} \varepsilon_{xx}^{a} \\ \varepsilon_{yy}^{a} \\ \varepsilon_{yz}^{a} \\ \varepsilon_{xz}^{a} \\ \varepsilon_{xy}^{a} \end{Bmatrix} = \begin{Bmatrix} \dfrac{\partial u_0}{\partial x} \\ \dfrac{\partial v_0}{\partial y} \\ \left(\phi_y + \dfrac{\partial w_0}{\partial y} \right) \\ \left(\phi_x + \dfrac{\partial w_0}{\partial x} \right) \\ \left(\dfrac{\partial u_0}{\partial y} + \dfrac{\partial v_0}{\partial x} \right) \end{Bmatrix} \begin{Bmatrix} \kappa_{xx}^{a} \\ \kappa_{yy}^{a} \\ \kappa_{yz}^{a} \\ \kappa_{xz}^{a} \\ \kappa_{xy}^{a} \end{Bmatrix} = \begin{Bmatrix} \dfrac{\partial \phi_x}{\partial x} \\ \dfrac{\partial \phi_y}{\partial y} \\ 0 \\ 0 \\ \dfrac{\partial \phi_x}{\partial y} + \dfrac{\partial \phi_y}{\partial x} \end{Bmatrix} \tag{4-7}$$

4.3 Governing Equation

The initial articulation of Hamilton's concept can be conveyed as Eq. (3–21) [6]:

4.3.1 Strain Energy

The preliminary fluctuation in the strain energy of the auxetic layer and magnetostrictive facesheets can be expressed as follows [7]:

$$\delta U^a = \int_0^a\int_0^b \Big(N_{xx}^a\left(\delta\varepsilon_{xx}^a\right) + M_{xx}^a\left(\delta\kappa_{xx}^a\right) + N_{yy}^a\left(\delta\varepsilon_{yy}^a\right) + M_{yy}^a\left(\delta\kappa_{yy}^a\right) + N_{xz}^a\left(\delta\varepsilon_{xz}^a\right)$$
$$+ M_{xz}^a\left(\delta\kappa_{xz}^a\right) + N_{yz}^a\left(\delta\varepsilon_{yz}^a\right) + M_{yz}^a\left(\delta\kappa_{yz}^a\right) + N_{xy}^a\left(\delta\varepsilon_{xy}^a\right) + M_{xy}^a\left(\delta\kappa_{xy}^a\right)\Big)dA \tag{4-8}$$

$$\delta U^m = \int_0^a\int_0^b \Big(N_{xx}^m\left(\delta\varepsilon_{xx}^m\right) + M_{xx}^m\left(\delta\kappa_{xx}^m\right) + N_{yy}^m\left(\delta\varepsilon_{yy}^m\right) + M_{yy}^m\left(\delta\kappa_{yy}^m\right) + N_{xz}^m\left(\delta\varepsilon_{xz}^m\right)$$
$$+ M_{xz}^m\left(\delta\kappa_{xz}^m\right) + N_{yz}^m\left(\delta\varepsilon_{yz}^m\right) + M_{yz}^m\left(\delta\kappa_{yz}^m\right) + N_{xy}^m\left(\delta\varepsilon_{xy}^m\right) + M_{xy}^m\left(\delta\kappa_{xy}^m\right)\Big)dA \tag{4-9}$$

in which,

$$\left(N_{ij}^{\ a}, M_{ij}^{\ a}, F_{ij}^{\ a}, G_{ij}^{\ a}\right) = \int_{-\frac{h_a}{2}}^{\frac{h_a}{2}} \sigma_{ij}^{\ a}\left(1, z, f(z), f(z)'\right)dz, \ (i,j = x,y,z)$$

$$\left(N_{ij}^{\ m}, M_{ij}^{\ m}, F_{ij}^{\ m}, G_{ij}^{\ m}\right) = \int_{\frac{h_a}{2}}^{\frac{h_a}{2}+h_m} \sigma_{ij}^{\ f}\left(1, z, f(z), f(z)'\right)dz$$
$$+ \int_{-\left(\frac{h_a}{2}+h_m\right)}^{-\frac{h_a}{2}} \sigma_{ij}^{\ m}\left(1, z, f(z), f(z)'\right)dz, \ (i,j = x,y,z) \tag{4-10}$$

The aforementioned equations utilize the superscripts "a" and "m" to denote the auxetic layer and magnetostrictive facesheets, respectively. The nonlocal correlations related to forces, moments, and stresses are denoted as [8, 9]:

$$\left(1-\mu^2\nabla^2\right)\begin{Bmatrix}\{N\}\\\{M\}\\\{F\}\end{Bmatrix} = \begin{bmatrix}[A] & [B] & [S]\\ [B] & [D] & [M]\\ [S] & [M] & [O]\end{bmatrix}\begin{Bmatrix}\{\varepsilon\}\\\{\kappa\}\\\{\chi\}\end{Bmatrix} \left(1-\mu^2\nabla^2\right)\begin{Bmatrix}G_{yz}\\G_{xz}\end{Bmatrix} = \begin{bmatrix}0 & L_{44}\\ L_{55} & 0\end{bmatrix}\begin{Bmatrix}X_{yz}\\X_{xz}\end{Bmatrix} \tag{4-11}$$

in which,

$$\{N\} = \{N_{xx} \quad N_{yy} \quad N_{xy}\}^T, \{M\} = \{M_{xx} \quad M_{yy} \quad M_{xy}\}^T$$
$$\{F\} = \{F_{xx} \quad F_{yy} \quad F_{xy}\}^T \{\in\} = \{\in_{xx} \in_{yy} \in_{xy}\}^T,$$
$$\{\kappa\} = \{\kappa_{xx} \quad \kappa_{yy} \quad \kappa_{xy}\}^T, \{\chi\} = \{\chi_{xx} \quad \chi_{yy} \quad \chi_{xy}\}^T \tag{4-12}$$

in which,

$$\begin{bmatrix} A_{11} & B_{11} & D_{11} & S_{11} & M_{11} & O_{11} & L_{11} \\ A_{22} & B_{22} & D_{22} & S_{22} & M_{22} & O_{22} & L_{22} \\ A_{12} & B_{12} & D_{12} & S_{12} & M_{12} & O_{12} & L_{12} \\ A_{44} & B_{44} & D_{44} & S_{44} & M_{44} & O_{44} & L_{44} \\ A_{55} & B_{55} & D_{55} & S_{55} & M_{55} & O_{55} & L_{55} \\ A_{66} & B_{66} & D_{66} & S_{66} & M_{66} & O_{66} & L_{66} \end{bmatrix} = \int_{-\frac{ha}{2}}^{\frac{ha}{2}} \begin{bmatrix} 1 \\ z \\ z^2 \\ f(z) \\ f(z)^2 \\ f(z)'^2 \end{bmatrix}^T \begin{bmatrix} C_{11}^a \\ C_{22}^a \\ C_{12}^a \\ C_{44}^a \\ C_{55}^a \\ C_{66}^a \end{bmatrix}$$
$$dz + \int_{\frac{ha}{2}}^{\frac{h_a}{2}+h_m} \begin{bmatrix} 1 \\ z \\ z^2 \\ f(z) \\ f(z)^2 \\ f(z)'^2 \end{bmatrix}^T \begin{bmatrix} C_{11}^m \\ C_{22}^m \\ C_{12}^m \\ C_{44}^m \\ C_{55}^m \\ C_{66}^m \end{bmatrix} dz + \int_{-\left(\frac{h_a}{2}+h_m\right)}^{-\frac{ha}{2}} \begin{bmatrix} 1 \\ z \\ z^2 \\ f(z) \\ f(z)^2 \\ f(z)'^2 \end{bmatrix}^T \begin{bmatrix} C_{11}^m \\ C_{22}^m \\ C_{12}^m \\ C_{44}^m \\ C_{55}^m \\ C_{66}^m \end{bmatrix} dz \tag{4-13}$$

4.3.2 External Work

The externally applied forces in this study arise from two factors:

- Winkler and Pasternak Foundation
- Uniformly distributed planar loads

Numerous significant interactions distinctly define work as a unique and particular physical quantity. Work is a form of energy transferred to or from the nanoplate due to the application of an external force. The device being examined utilizes a sandwich nanoplate composed of magnetostrictive and auxetic materials. The term "work" refers to the energy applied by external forces along the axis of the nanoplate, resulting in displacement and deformation within the nanoplate.

The initial version of the exterior work can be articulated as [10–12]:

$$\delta V = \int_0^a \int_0^b \left[k_w . \delta w_0 - kg\left(\frac{\partial^2 \delta w_0}{\partial x^2} + \frac{\partial^2 \delta w_0}{\partial y^2} \right) - N_{xx}\left(\frac{\partial^2 \delta w_0}{\partial x^2} \right) - N_{yy}\left(\frac{\partial^2 \delta w_0}{\partial y^2} \right) \right] \tag{4–14}$$

The values k_w and k_P denote the Winkler stiffness and shear layer stiffness, respectively. The dimensionless representation of the Winkler and Pasternak foundation is defined as follows:

$$K_w = \frac{k_w a^4}{D_{11}} \quad K_g = \frac{k_g a^2}{D_{11}}$$

The pre-buckling resultant forces are depicted in Figure 4.2:

$$N_{xx} = \gamma_1 P, N_{yy} = \gamma_2 P \tag{4-15}$$

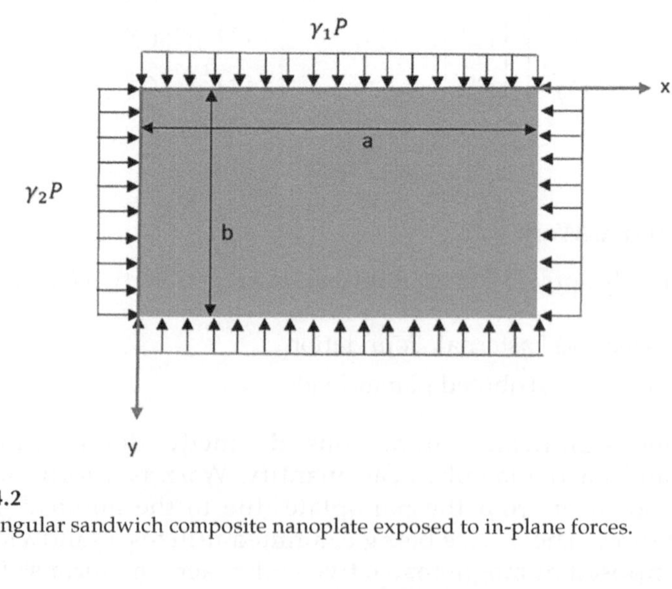

FIGURE 4.2
The rectangular sandwich composite nanoplate exposed to in-plane forces.

4.3.3 Kinetic Energy

The first expression of the kinetic energy equation can be denoted as follows:

$$\delta T=\frac{1}{2}\iint\left[-I_0\left(\frac{\partial^2 u_0}{\partial t^2}\right)\delta u_0-I_1\left(\frac{\partial^3 u_0}{\partial x\partial t^2}\right)\delta w_0-I_3\left(\frac{\partial^2 u_0}{\partial t^2}\right)\delta\phi_x+\right.$$
$$I_1\left(\frac{\partial^3 w_0}{\partial x\partial t^2}\right)\delta u_0+I_2\left(\frac{\partial^4 w_0}{\partial x^2\partial t^2}\right)\delta w_0+I_4\left(\frac{\partial^3 w_0}{\partial x\partial t^2}\right)\delta\phi_x-I_3\left(\frac{\partial^2 \phi_x}{\partial t^2}\right)\delta u_0-$$
$$I_4\left(\frac{\partial^3 \phi_x}{\partial x\partial t^2}\right)\delta w_0-I_5\left(\frac{\partial^2 \phi_x}{\partial t^2}\right)\delta\phi_x-I_0\left(\frac{\partial^2 v_0}{\partial t^2}\right)\delta v_0-I_1\left(\frac{\partial^3 v_0}{\partial y\partial t^2}\right)\delta w_0-$$
$$I_3\left(\frac{\partial^2 v_0}{\partial t^2}\right)\delta\phi_y+I_1\left(\frac{\partial^3 w_0}{\partial y\partial t^2}\right)\delta v_0+I_2\left(\frac{\partial^4 w_0}{\partial y^2\partial t^2}\right)\delta w_0+I_4\left(\frac{\partial^3 w_0}{\partial y\partial t^2}\right)\delta\phi_y-$$
$$\left.I_3\left(\frac{\partial^2 \phi_y}{\partial t^2}\right)\delta v_0-I_4\left(\frac{\partial^3 \phi_y}{\partial y\partial t^2}\right)\delta w_0-I_5\left(\frac{\partial^2 \phi_y}{\partial t^2}\right)\delta\phi_y-I_0\left(\frac{\partial^2 w_0}{\partial t^2}\right)\delta w_0\right]dA \quad (4\text{-}16)$$

The mass moment of inertia is defined as follows:

$$I_i=\left(\int_{-h_a/2}^{h_a/2}\Upsilon_i\rho_a dz\right)+\left(\int_{\frac{(h_a+h_m)}{2}}^{\frac{-h_a}{2}}\Upsilon_i\rho_f dz+\int_{\frac{ha}{2}}^{\frac{(h_a+h_m)}{2}}\Upsilon_i\rho_f dz\right),$$
$$\Upsilon_i=\left[1,z,z^2,f(z),zf(z),f(z)^2\right],i=(1,2,3,4,5) \quad (4\text{-}17)$$

4.4 Equation of Motion

The Euler-Lagrange equation for sinusoidal shear deformation integrates the initial change in strain energy, the work done by applied forces, and the kinetic energy to formulate the five equilibrium equations. The stated result is achieved by substituting Eqs. (4–8) to (4–17) with Eq. (3–21), resulting in:

$$\boldsymbol{\delta u_0}:\frac{\partial N_{xx}}{\partial x}+\frac{\partial N_{xy}}{\partial y}-e_{31}h_cK_cC(t)\frac{\partial^2 w_0}{\partial x\partial t}=I_0\left(\frac{\partial^2 u_0}{\partial t^2}\right)+I_3\left(\frac{\partial^2 \phi_x}{\partial t^2}\right)-I_1\left(\frac{\partial^3 w_0}{\partial x\partial t^2}\right)$$
$$(4\text{-}18a)$$

$$\delta v_0: \frac{\partial N_{xx}}{\partial y} + \frac{\partial N_{xy}}{\partial x} - e_{32} h_c K_c C(t) \frac{\partial^2 w_0}{\partial y \partial t}$$

$$= I_0 \left(\frac{\partial^2 v_0}{\partial t^2} \right) + I_3 \left(\frac{\partial^2 \phi_y}{\partial t^2} \right) - I_1 \left(\frac{\partial^3 w_0}{\partial y \partial t^2} \right) \quad (4\text{–}18b)$$

$$\delta w_0: \frac{\partial^2 M_{xx}}{\partial x^2} + \frac{\partial^2 M_{yy}}{\partial y^2} + 2\frac{\partial^2 M_{xy}}{\partial x \partial y} - k_w . w_0 + k_g \left(\frac{\partial^2 w_0}{\partial x^2} + \frac{\partial^2 w_0}{\partial y^2} \right)$$

$$-N_{xx} \left(\frac{\partial^2 w_0}{\partial x^2} \right) - N_{yy} \left(\frac{\partial^2 w_0}{\partial y^2} \right) - e_{31} h_c K_c C(t) \frac{\partial^2 U_x}{\partial x \partial t}$$

$$-e_{32} h_c K_c C(t) \frac{\partial^2 V_y}{\partial y \partial t} = I_0 \left(\frac{\partial^2 w_0}{\partial t^2} \right) + I_1 \left(\frac{\partial^3 u_0}{\partial x \partial t^2} \right) + I_1 \left(\frac{\partial^3 v_0}{\partial y \partial t^2} \right)$$

$$-I_2 \left(\frac{\partial^4 w_0}{\partial x^2 \partial t^2} \right) - I_2 \left(\frac{\partial^4 w_0}{\partial y^2 \partial t^2} \right) + I_4 \left(\frac{\partial^3 \phi_x}{\partial x \partial t^2} \right) + I_4 \left(\frac{\partial^3 \phi_y}{\partial y \partial t^2} \right) \quad (4\text{–}18c)$$

$$\delta \phi_x: \frac{\partial F_{xx}}{\partial x} + \frac{\partial F_{xy}}{\partial y} - G_{xz} = I_3 \left(\frac{\partial^2 u_0}{\partial t^2} \right) + I_5 \left(\frac{\partial^2 \phi_x}{\partial t^2} \right) - I_4 \left(\frac{\partial^3 w_0}{\partial x \partial t^2} \right) \quad (4\text{–}18d)$$

$$\delta \phi_y: \frac{\partial F_{yy}}{\partial y} + \frac{\partial F_{xy}}{\partial x} - G_{yz} = I_3 \left(\frac{\partial^2 v_0}{\partial t^2} \right) + I_5 \left(\frac{\partial^2 \phi_y}{\partial t^2} \right) - I_4 \left(\frac{\partial^3 w_0}{\partial y \partial t^2} \right) \quad (4\text{–}18e)$$

4.5 Precise Solution for the Buckling of a Nanoplate

In the Galerkin approach, it is crucial that the offered solutions conform to the specified boundary conditions of the issue. The computation of the critical buckling loads will be conducted by substituting the specified response, as indicated in Eq. (4–19), into the governing equation. The Galerkin method is extensively employed in the study of buckling processes and is a dependable methodology that effectively resolves partial differential equations with significant precision, convergence, and computational efficiency. The needed sequence of displacement components to meet the aforementioned boundary criteria is delineated in the following formats:

$$\begin{Bmatrix} u_0(x,y,t) \\ v_0(x,y,t) \\ w_0(x,y,t) \\ \phi_x(x,y,t) \\ \phi_y(x,y,t) \end{Bmatrix} = \begin{Bmatrix} \sum_{m=1}^{\infty}\sum_{n=1}^{\infty} U_{mn} \frac{\partial X_m}{\partial x} Y_n e^{i\omega t} \\ \sum_{m=1}^{\infty}\sum_{n=1}^{\infty} V_{mn} X_m \frac{\partial Y_n}{\partial y} e^{i\omega t} \\ \sum_{m=1}^{\infty}\sum_{n=1}^{\infty} W_{mn} X_m Y_n e^{i\omega t} \\ \sum_{m=1}^{\infty}\sum_{n=1}^{\infty} \phi_{xmn} \frac{\partial X_m}{\partial x} Y_n e^{i\omega t} \\ \sum_{m=1}^{\infty}\sum_{n=1}^{\infty} \phi_{ymn} X_m \frac{\partial Y_n}{\partial y} e^{i\omega t} \end{Bmatrix} \tag{4-19}$$

The vibration amplitudes are represented by U_{mn}, V_{mn}, W_{mn}, ϕ_{xmn}, and ϕ_{ymn}. The functions X_m $X_m(x)$ and $Y_n(y)$ are deemed appropriate as they are defined according to different boundary conditions, as illustrated in Table 4.1.

By isolating the variables $X_m(a) = X^{//}{}_m(a)$ in Eqs (4–18a)–(4–18e), the natural frequencies can be ascertained by solving the resultant equation. To get the natural frequencies, it is essential to equate the determinant of the following equation to zero.

$$\left| [K] - \omega^2 [M] + i\omega [C] \right| = 0 \tag{4-20}$$

The matrices K, M, and C denote the stiffness, mass, and damping matrices, respectively.

4.5.1 Quantitative Findings

This section will clarify the chosen findings of the current research, focusing on a thorough analysis of the impact of different parameters on the non-dimensional critical buckling load. To provide a graphical representation, the parameter denoting the magnitude of the magnetostrictive layer, commonly known as the core layer, is represented as h_a, while h_m denotes the dimensions of the facesheets.

4.5.2 Findings and Analysis

Table 4.1 has been created to examine the effects of the auxetic inclination angle and velocity feedback gain on the dimensionless buckling load. The data presented in Table 4.2 indicate that increasing the auxetic inclination

TABLE 4.1
The Permissible Functions X_m (x) and Y_n (y) [2]

	Boundary Conditions		The Function X_m and Y_n	
	At x = 0, a	At y = 0, b	$X_m(x)$	$Y_n(y)$
SSSS	$X_m(a) = X''_m(0) = 0$	$Y_n(0) = Y''_n(0) = 0$	$\sin(\alpha x)$	$\sin(\beta y)$
	$X_m(a) = X''_m(a) = 0$	$Y_n(b) = Y''_n(b) = 0$		
CCSS	$X_m(0) = X'_m(0) = 0$	$Y_n(0) = Y''_n(0) = 0$	$\sin^2(\alpha x)$	$\sin(\beta y)$
	$X_m(a) = X'_m(a) = 0$	$Y_n(b) = Y''_n(b) = 0$		
CCCC	$X_m(0) = X'_m(0) = 0$	$Y_n(0) = Y'_n(0) = 0$	$\sin^2(\alpha x)$	$\sin^2(\beta y)$
	$X_m(a) = X'_m(0) = 0$	$Y_n(b) = Y'_n(b) = 0$		
CCFF	$X''_m(0) = X'''_m(0) = 0$	$Y_n(0) = Y'_n(0) = 0$	$\cos^2(\alpha x)\left[\sin^2(\alpha x) + 1\right]$	$\sin^2(\beta y)$
	$X''_m(a) = X'''_m(a) = 0$	$Y_n(b) = Y'_n(b) = 0$		

TABLE 4.2

Dimensionless Buckling Load of a Square Plate for Various Auxetic Inclination Angles and Magnetostrictive Thickness (t = 1nm, l = 4nm, e = 8nm, a = b = 40nm, ha = 4nm, hm = 0.5nm, SSSS, γ_1 = -1, γ_2=0)

	$\theta = -75°$	$\theta = -60°$	$\theta = -30°$	$\theta = -15°$	$\theta = +15°$	$\theta = +30°$	$\theta = +60°$	$\theta = +75°$
$h_m = 0.5nm$	7.74098	7.73921	7.73604	7.736	7.73699	7.73786	7.73724	7.73607
$h_m = 0.25nm$	3.44445	3.44269	3.43951	3.43947	3.44046	3.44133	3.44072	3.43955
$h_m = 0.15nm$	1.97089	1.96912	1.96594	1.96591	1.96689	1.96777	1.96715	1.96598
$h_m = 0.05nm$	0.628214	0.626448	0.623271	0.623236	0.624222	0.625094	0.624477	0.623309

angle results in a reduction of the dimensionless buckling stress. This decline may be ascribed to the decrease in flexural stiffness. This behavior can primarily be ascribed to the values of Poisson's ratio. Moreover, as illustrated in Table 4.2, the enhancement of the magnetostrictive layer results in a reduction of the dimensionless buckling load of the system.

Figure 4.3 illustrates a notable decrease in dimensionless buckling with an increase in the inclination angle. This observation corresponds with the reasoning articulated in the preceding paragraphs from a physics standpoint. Additionally, Figure 4.3 depicts the influence of the magnetostrictive thickness parameter on the dimensionless buckling load. Figure 4.3 illustrates a notable increase in the buckling load with the augmentation of magnetostrictive thickness.

Tables 4.3 and 4.4 have been provided to examine the influence of vertical rib length and thickness of auxetic materials on the dimensionless buckling load, taking into account different values of velocity feedback gain. The data unequivocally indicate that an augmentation in vertical rib length correlates with an increase in the dimensionless buckling stress. The data presented in this table indicate that an increase in feedback gain corresponds to a decline in the dimensionless buckling load.

The influence of vertical rib length on the buckling load of the system has been examined under both SSSS and SSCC boundary conditions, as illustrated in Figure 4.4. Figure 4.4 illustrates a clear association between the augmentation of vertical rib length and the subsequent enhancement in

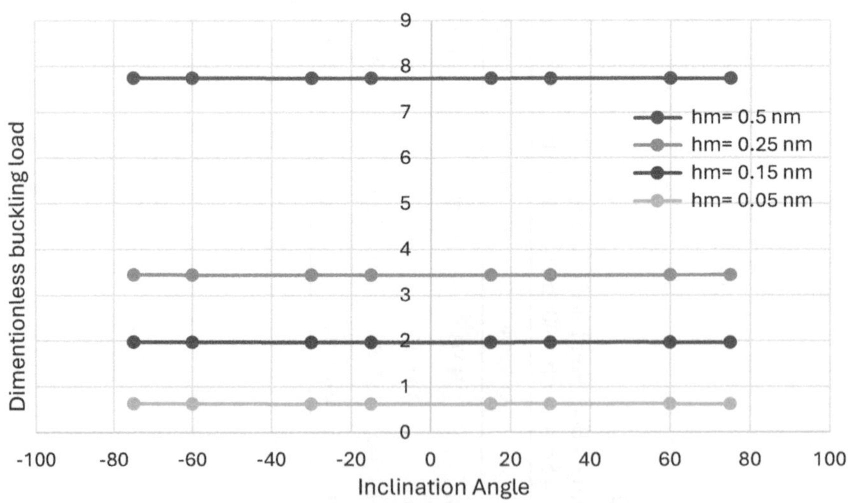

FIGURE 4.3
Dimensionless buckling load as a function of inclination angle (t = 1nm, l = 4nm, e = 8nm, a = b = 40nm, ha = 4nm, hm = 0.5nm, SSSS, γ_1 = -1, γ_2 = 0).

TABLE 4.3

Dimensionless Buckling Load of a Square Plate for Various Rib Lengths and Nonlocal Parameters (t = 0.25 nm, e = 8 nm, a = b = 40 nm, ha = 4 nm, hm = 0.1 nm, θ = +60°, γ_1 = -1, γ_2=0)

		$l = 8nm$	$l = 4nm$	$l = 2nm$	$l = 1nm$	$l = 0.5nm$
$\mu^2 = 0\ nm^2$	SSSS	1.27951	1.27951	1.27918	1.27861	1.27426
$\mu^2 = 1\ nm^2$		1.26392	1.26392	1.26359	1.26303	1.25873
$\mu^2 = 4\ nm^2$		1.2487	1.2487	1.24838	1.24782	1.24357
$\mu^2 = 0\ nm^2$	SSCC	2.15917	2.15919	2.15866	2.15798	2.1529
$\mu^2 = 1\ nm^2$		2.13286	2.13287	2.13235	2.13168	2.12667
$\mu^2 = 4\ nm^2$		2.10718	2.10719	2.10668	2.10601	2.10106

TABLE 4.4

Dimensionless Buckling Load of a Square Plate for Various Auxetic Thickness Values and Nonlocal Parameters (l = 2nm, e = 8nm, a = b = 40nm, ha = 4nm, hm = 0.1nm, θ = +60°, γ_1 = -1, γ_2=0)

		t=1nm	t=0.5nm	t=0.25nm	t=0.1nm	t=0.01nm
$\mu^2 = 0\ nm^2$	SSSS	1.25847	1.27688	1.27918	1.27949	1.27951
$\mu^2 = 1\ nm^2$		1.24313	1.26132	1.26359	1.2639	1.26392
$\mu^2 = 4\ nm^2$		1.22817	1.24613	1.24838	1.24868	1.2487
$\mu^2 = 0\ nm^2$	SSCC	2.12626	2.15506	2.15866	2.15914	2.15917
$\mu^2 = 1\ nm^2$		2.10035	2.12879	2.13235	2.13283	2.13286
$\mu^2 = 4\ nm^2$		2.07506	2.10316	2.10668	2.10715	2.10718

the dimensionless buckling stress. Furthermore, as the nonlocal parameter increases, the dimensionless buckling stress of the system diminishes.

Table 4.5 displays the non-dimensional buckling stress for several combinations of aspect ratio (a/b), side-to-thickness ratio (a/h_a), and inclination angle. The table illustrates a positive association between the aspect ratio (a/b) and the non-dimensional buckling load, signifying that an increase in the aspect ratio results in a proportional rise in the non-dimensional buckling load. Furthermore, the non-dimensional buckling stress increases with an increase in the side-to-thickness ratio (a/h_a).

Table 4.6 presents a comparative analysis of the non-dimensional buckling load under various external loads. Furthermore, it is expected that

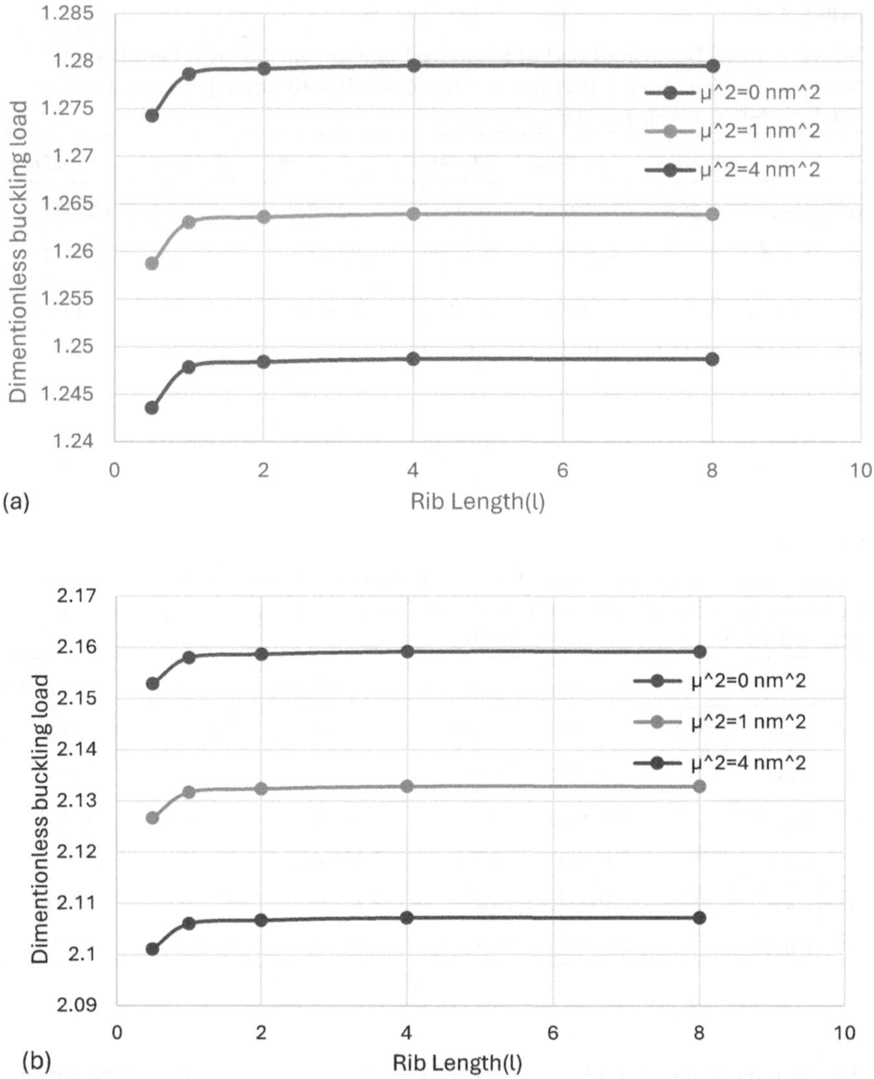

FIGURE 4.4
Dimensionless buckling load as a function of vertical rib length (a) SSSS and (b) SSCC.

the value of the Winkler and Pasternak foundation will remain stable when Kw equals zero and Kg equals zero.

Figure 4.5 illustrates the influence of thickness on the performance of the continuous system, particularly under the SSSS and SSCC boundary conditions. An increase in thickness (t) results in a decrease in the non-dimensional buckling stress. Moreover, it has been demonstrated that the

TABLE 4.5

Non-Dimensional Buckling Load for Various Aspect Ratios under SSSS Boundary Conditions (μ^2 = 0 nm^2, K_w = K_g = 0, a = 40 nm, hm = 0.5 nm, γ_1 = -1, γ_2 = 0)

a/b	θ	a/h_a			
		5	10	20	30
0.5	−75°	2.69109	6.04775	15.0649	27.6493
1		3.44445	7.74098	3.4445	35.391
0.5	−60°	2.68947	6.04613	15.0633	27.6477
1		3.44269	7.73921	3.44269	35.3892
0.5	−30°	2.68737	6.04403	15.0612	27.6456
1		3.43951	7.73604	3.43951	35.386
0.5	−15°	2.68717	6.0483	15.061	27.6454
1		3.43947	7.736	3.43947	35.386

TABLE 4.6

Comparison of the Non-Dimensional Buckling Load for SSSS And SSCC Boundary Conditions (μ^2 = 1 nm², K_w = K_g = 0, t = 0.25 nm, l = 2 nm, e = 7 nm, ha = 3 nm, hm = 0.1 nm)

θ°	γ_1, γ_2	SSSS	SSCC
60°	(-1, 0)	1.73324	2.88923
	(-1, -1)	0.856057	1.44462
30°	(-1, 0)	1.73424	2.89078
	(-1, -1)	0.856551	1.44539

buckling load of the structure significantly decreases when the thickness (t) parameter approaches 0.5 nm.

4.5.3 Final Assessment

- The dimensional buckling load exhibits a modest decrease with the elongation of the rib's length.
- The confinement of the dimensional buckling load with simply clamped boundary conditions exceeds that of the dimensionless buckling load under simple boundary conditions.
- The dimensionless buckling load decreases with an increase in the thickness of the magnetostrictive layer.
- Augmenting the thickness (t) parameter leads to a reduction in the critical buckling load.

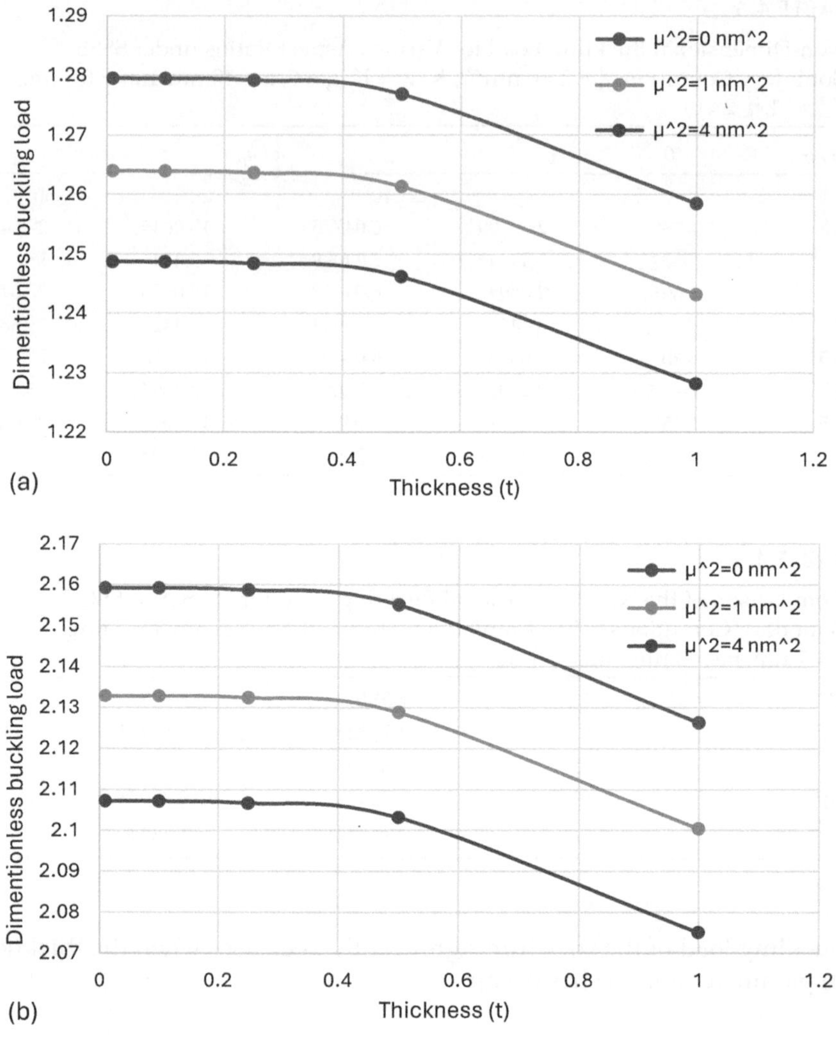

FIGURE 4.5
Dimensionless buckling load as a function of thickness (t) for (a) SSSS and (b) SSCC.

4.6 Buckling of Meta-Graphene-Origami-Enabled Magnetostrictive Nanoplates Subjected to Temperature Gradient

The new field of nano-architected materials has reached a new level with the evolution of multi-functional systems that exhibit unprecedented mechanical and adaptive behavior. This section investigates the buckling response of a new class of these materials: meta-graphene-origami substrate-engineered magnetostrictive nanoplates. Unlike classical structures, this structure exploits the tunable porosity and negative Poisson's ratio of meta-origami lattices, the exceptional electrical conductivity and strength of graphene, and the inherent strain-magnetic field coupling of magnetostrictive materials. Under the action of a thermal gradient, this multi-physics system experiences intricate thermomechanical stresses that are in essence reshaped by the multi-physics interactions within its constituents. Critical buckling load under such conditions is most critical to study since it dictates the operational limits and functional reliability of such nanoplates in future devices such as nano-electromechanical systems (NEMS), drug delivery, and adaptive meta-surfaces for heat control.

4.7 Background

Origami-inspired structural ideas offer programmable mechanical properties in the form of folding-induced stiffness modulation and enhanced structural resilience. When applied to graphene-based nanostructures, these features deliver tunable folding pattern control over instability mechanisms, improved load-bearing and energy dissipation capability, and coupled mechanical and thermal and magnetic field sensitivity toward intelligent functions.

The combination of graphene origami and magnetostrictive materials provides dynamic control of buckling via external magnetic fields, producing reconfigurable and deformable changes. Temperature gradients play a crucial role in influencing the mechanical response of nanoplates by inducing thermal stresses, altering material properties, and modifying magnetostrictive effects.

4.8 Theoretical Framework and Formulation

This research examines a rectangular plate characterized by its dimensions: length (a), width (b), and total thickness (h) along the x, y, and z axes, respectively. The plate consists of nano-fac sheets exhibiting magnetostrictive properties, along with a core layer that displays behavior permitted by GOA. The present investigation used the Cartesian coordinate system, illustrated in Figure 4.6, with the origin positioned at one corner of the plate. The methodology established by Winkler and Pasternak has been employed in the examination of phenomena associated with the foundation effect.

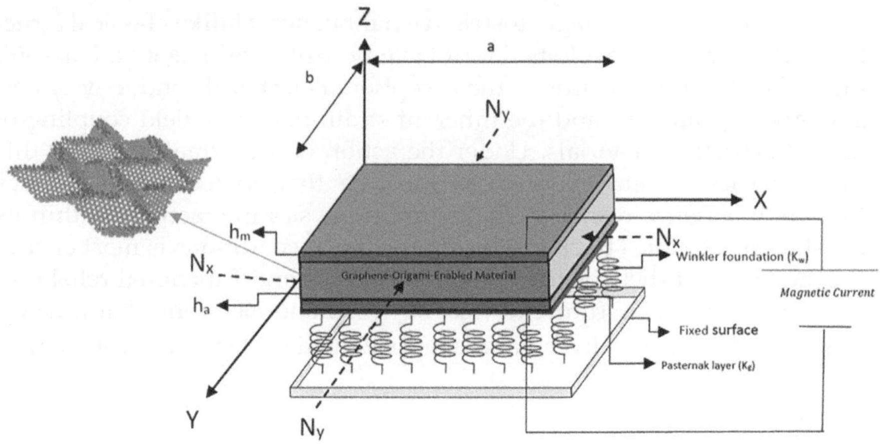

FIGURE 4.6
Arrangement of sandwich composite.

4.9 Materials Enabled by Graphene Origami

The derivation approach will yield the characteristics of graphene-origami auxetic (GOA):

$$E^{Auxetic} = E_{Cu} \frac{1+\xi\eta V_{Gr}}{1-\eta V_{Gr}} f_E\left(H_{Gr}, V_{Gr}, T\right) \tag{4-21}$$

$$\nu^{Auxetic} = \left(V_{Gr}\nu_{Gr} + V_{Cu}\nu_{Cu}\right) f_v\left(H_{Gr}, V_{Gr}, T\right) \tag{4-22}$$

$$\alpha^{Auxetic} = \left(V_{Gr}\alpha_{Gr} + V_{Cu}\alpha_{Cu}\right) f_\alpha\left(V_{Gr}, T\right) \tag{4-23}$$

$$\rho^{Auxetic} = \left(V_{Gr}\rho_{Gr} + V_{Cu}\rho_{Cu}\right) f_\rho\left(V_{Gr}, T\right) \tag{4-24}$$

$$f_E\left(H_{Gr}, V_{Gr}, T\right) = \begin{pmatrix} 1.11 - 1.22V_{Gr} - 0.134\left(\frac{T}{T_0}\right) \\ +0.559V_{Gr}\left(\frac{T}{T_0}\right) - 5.5H_{Gr}V_{Gr} \\ +38H_{Gr}V_{Gr}^2 - 20.6H_{Gr}^2V_{Gr}^2 \end{pmatrix} \tag{4-25}$$

$$f_v\left(H_{Gr},V_{Gr},T\right)=\begin{pmatrix} 1.01-1.43V_{Gr}+0.165\left(\dfrac{T}{T_0}\right) \\ -16.8H_{Gr}V_{Gr}-1.1H_{Gr}V_{Gr}\left(\dfrac{T}{T_0}\right) \\ +16H_{Gr}^2V_{Gr}^2 \end{pmatrix} \tag{4-26}$$

$$f_\alpha\left(V_{Gr},T\right)=\begin{pmatrix} 0.794-16.8V_{Gr}^2-0.0279\left(\dfrac{T}{T_0}\right)^2 \\ +0.182\left(\dfrac{T}{T_0}\right)\left(V_{Gr}+1\right) \end{pmatrix} \tag{4-27}$$

$$f_\rho\left(V_{Gr},T\right)=\left(1.01-2.01V_{Gr}^2-0.0131\left(\frac{T}{T_0}\right)\right) \tag{4-28}$$

The temperature T_0 is 300 Kelvin, while T is the ambient temperature. The volume proportion $\left(V_{Gr}\right)$ of the graphene-origami auxetic material for uniform distribution is officially stated as follows:

$$V_{Gr}=\frac{\rho_{Cu}W_{Gr}}{\rho_{Cu}W_{Gr}+\rho_{Gr}\left(1-W_{Gr}\right)}\quad V_{Cu}=1-V_{Gr} \tag{4-29}$$

The characters ρ_{Cu} and ρ_{Gr} denote the respective densities of pure copper and graphene. The sign V_{Cu} denotes the volume percentage of copper. Furthermore, it is posited that the extent of folding in graphene-origami, as determined by the hydrogen atom coverage $\left(H_{Gr}\right)$ at the crease, exhibits a progressive variation along the thickness direction. The material coefficient (η) and size coefficient (ξ) are clearly defined as follows:

$$\eta=\frac{\dfrac{E_{Gr}}{E_{Cu}}-1}{\dfrac{E_{Gr}}{E_{Cu}}+\xi},\ \xi=2\left(\frac{l_{Gr}}{t_{Gr}}\right) \tag{4-30}$$

The values l_{Gr} and t_{Gr} denote the length and thickness of graphene, respectively.

4.10 Geometric Configuration

This study employs the higher-order sinusoidal shear deformation plate theory originally introduced by Touratier [1]. According to the principles of higher-order sinusoidal shear deformation plate theory, the neutral surface, aligned normally to the midline, maintains its perpendicular orientation both prior to and following deformation.

4.10.1 Displacement Field

The derivation of the displacement field can be achieved by examining the postulates of the higher-order sinusoidal shear deformation plate theory [2] as Eq. (4–1) to Eq. (4–3).

4.11 Eringen's Nonlocal Theory

According to Eringen's nonlocal theory, commonly applied in micro and nanoscale scenarios, the stress at a certain reference point (x) inside an elastic continuum is influenced not only by the strain at point x but also by the strains at all adjacent points. The previously noted link can be articulated as Eq. (3–9).

4.12 Stress-Strain Relationship for Magnetostrictive Materials

The unique properties exhibited by magnetostrictive materials have attracted considerable interest in the engineering field due to their ability to convert energy among several forms, including magnetic, electric, and mechanical. The equation denotes the stress, strain, and magnetic susceptibility of the magnetostrictive layer as Eqs. (3–43) to (3–47).

4.12.1 Velocity Feedback Coefficient

The magnetic field $\left(H_{zz}\right)$ can be calculated using Eq. (3–13). The non-zero strains shown by the magnetostrictive layer can be defined as [2]:

$$\begin{Bmatrix} \varepsilon_{xx}^{m} \\ \varepsilon_{yy}^{m} \\ \varepsilon_{yz}^{m} \\ \varepsilon_{xz}^{m} \\ \varepsilon_{xy}^{m} \end{Bmatrix} = \begin{Bmatrix} \varepsilon_{xx}^{m} \\ \varepsilon_{yy}^{m} \\ \varepsilon_{yz}^{m} \\ \varepsilon_{xz}^{m} \\ \varepsilon_{xy}^{m} \end{Bmatrix} + z \begin{Bmatrix} \kappa_{xx}^{m} \\ \kappa_{yy}^{m} \\ \kappa_{yz}^{m} \\ \kappa_{xz}^{m} \\ \kappa_{xy}^{m} \end{Bmatrix} + f(z) \begin{Bmatrix} \chi_{xx}^{m} \\ \chi_{yy}^{m} \\ \chi_{yz}^{m} \\ \chi_{xz}^{m} \\ \chi_{xy}^{m} \end{Bmatrix} + f(z)' \begin{Bmatrix} X_{xx}^{m} \\ X_{yy}^{m} \\ X_{yz}^{m} \\ X_{xz}^{m} \\ X_{xy}^{m} \end{Bmatrix} \tag{4-31}$$

where,

$$\begin{Bmatrix} \varepsilon_{xx}^{m} \\ \varepsilon_{yy}^{m} \\ \varepsilon_{yz}^{m} \\ \varepsilon_{xz}^{m} \\ \varepsilon_{xy}^{m} \end{Bmatrix} = \begin{Bmatrix} \dfrac{\partial u_0}{\partial x} \\ \dfrac{\partial v_0}{\partial y} \\ 0 \\ 0 \\ \left(\dfrac{\partial u_0}{\partial y} + \dfrac{\partial v_0}{\partial x}\right) \end{Bmatrix}, \begin{Bmatrix} \kappa_{xx}^{m} \\ \kappa_{yy}^{m} \\ \kappa_{yz}^{m} \\ \kappa_{xz}^{m} \\ \kappa_{xy}^{m} \end{Bmatrix} = \begin{Bmatrix} -\dfrac{\partial^2 w_0}{\partial x^2} \\ -\dfrac{\partial^2 w_0}{\partial y^2} \\ 0 \\ 0 \\ -\left(\dfrac{\partial^2 w_0}{\partial x^2} + \dfrac{\partial^2 w_0}{\partial y^2}\right) \end{Bmatrix}, \begin{Bmatrix} \chi_{xx}^{m} \\ \chi_{yy}^{m} \\ \chi_{yz}^{m} \\ \chi_{xz}^{m} \\ \chi_{xy}^{m} \end{Bmatrix} = \begin{Bmatrix} \dfrac{\partial \phi_x}{\partial x} \\ \dfrac{\partial \phi_y}{\partial y} \\ 0 \\ 0 \\ \left(\dfrac{\partial \phi_x}{\partial y} + \dfrac{\partial \phi_y}{\partial x}\right) \end{Bmatrix} \tag{4-32}$$

4.12.2 Sequential Stress-Strain Relationship of Graphene-Origami-Enhanced Materials

In the domain of graphene-origami materials, the correlation between stress and strain can be expressed as follows [5]:

$$\left(1-\mu^2\nabla^2\right) \begin{Bmatrix} \sigma_{xx}^{a,nl} \\ \sigma_{yy}^{a,nl} \\ \tau_{yz}^{a,nl} \\ \tau_{xz}^{a,nl} \\ \tau_{xy}^{a,nl} \end{Bmatrix} = \begin{bmatrix} C_{11}^{a} & C_{12}^{a} & 0 & 0 & 0 \\ C_{21}^{a} & C_{22}^{a} & 0 & 0 & 0 \\ 0 & 0 & C_{44}^{a} & 0 & 0 \\ 0 & 0 & 0 & C_{55}^{a} & 0 \\ 0 & 0 & 0 & 0 & C_{66}^{a} \end{bmatrix} \begin{Bmatrix} \varepsilon_{xx}^{a} \\ \varepsilon_{yy}^{a} \\ \gamma_{yz}^{a} \\ \gamma_{xz}^{a} \\ \gamma_{xy}^{a} \end{Bmatrix} \tag{4-33}$$

The non-zero stresses of a layer facilitated by graphene origami are delineated as follows:

$$\begin{Bmatrix} \varepsilon_{xx}^{a} \\ \varepsilon_{yy}^{a} \\ \varepsilon_{yz}^{a} \\ \varepsilon_{xz}^{a} \\ \varepsilon_{xy}^{a} \end{Bmatrix} = \begin{Bmatrix} \varepsilon_{xx}^{a} \\ \varepsilon_{yy}^{a} \\ \varepsilon_{yz}^{a} \\ \varepsilon_{xz}^{a} \\ \varepsilon_{xy}^{a} \end{Bmatrix} + z \begin{Bmatrix} \kappa_{xx}^{a} \\ \kappa_{yy}^{a} \\ \kappa_{yz}^{a} \\ \kappa_{xz}^{a} \\ \kappa_{xy}^{a} \end{Bmatrix} + f(z) \begin{Bmatrix} \chi_{xx}^{a} \\ \chi_{yy}^{a} \\ \chi_{yz}^{a} \\ \chi_{xz}^{a} \\ \chi_{xy}^{a} \end{Bmatrix} + f(z)' \begin{Bmatrix} X_{xx}^{a} \\ X_{yy}^{a} \\ X_{yz}^{a} \\ X_{xz}^{a} \\ X_{xy}^{a} \end{Bmatrix} \tag{4-34}$$

in which,

$$\begin{Bmatrix} \varepsilon_{xx}^{a} \\ \varepsilon_{yy}^{a} \\ \varepsilon_{yz}^{a} \\ \varepsilon_{xz}^{a} \\ \varepsilon_{xy}^{a} \end{Bmatrix} = \begin{Bmatrix} \frac{\partial u_0}{\partial x} \\ \frac{\partial v_0}{\partial y} \\ 0 \\ 0 \\ \left(\frac{\partial u_0}{\partial y} + \frac{\partial v_0}{\partial x} \right) \end{Bmatrix}, \begin{Bmatrix} \kappa_{xx}^{a} \\ \kappa_{yy}^{a} \\ \kappa_{yz}^{a} \\ \kappa_{xz}^{a} \\ \kappa_{xy}^{a} \end{Bmatrix} = \begin{Bmatrix} -\frac{\partial^2 w_0}{\partial x^2} \\ -\frac{\partial^2 w_0}{\partial y^2} \\ 0 \\ 0 \\ -\left(\frac{\partial^2 w_0}{\partial x^2} + \frac{\partial^2 w_0}{\partial y^2} \right) \end{Bmatrix}, \begin{Bmatrix} \chi_{xx}^{a} \\ \chi_{yy}^{a} \\ \chi_{yz}^{a} \\ \chi_{xz}^{a} \\ \chi_{xy}^{a} \end{Bmatrix} =$$

$$\begin{Bmatrix} \frac{\partial \phi_x}{\partial x} \\ \frac{\partial \phi_y}{\partial y} \\ 0 \\ 0 \\ \left(\frac{\partial \phi_x}{\partial y} + \frac{\partial \phi_y}{\partial x} \right) \end{Bmatrix}, \begin{Bmatrix} X_{xx}^{a} \\ X_{yy}^{a} \\ X_{yz}^{a} \\ X_{xz}^{a} \\ X_{xy}^{a} \end{Bmatrix} = \begin{Bmatrix} 0 \\ 0 \\ \phi_x \\ \phi_y \\ 0 \end{Bmatrix} \tag{4-35}$$

The characteristics of magnetostrictive layer graphene-origami-enabled auxetic materials are detailed in the cited references.

4.13 Regulatory Equation

The initial variation of Hamilton can be expressed as Eq. (3–21).

4.13.1 Strain Energy

The initial fluctuation in the strain energy of the layer facilitated by graphene-origami and the facesheets with magnetostrictive characteristics can be articulated as follows [56]:

$$\delta U^a = \int_0^a\int_0^b \Big(N_{xx}^a\left(\delta\varepsilon_{xx}^a\right) + M_{xx}^a\left(\delta\kappa_{xx}^a\right) + N_{yy}^a\left(\delta\varepsilon_{yy}^a\right) + M_{yy}^a\left(\delta\kappa_{yy}^a\right) + N_{xz}^a\left(\delta\varepsilon_{xz}^a\right) + M_{xz}^a\left(\delta\kappa_{xz}^a\right) + N_{yz}^a\left(\delta\varepsilon_{yz}^a\right) + M_{yz}^a\left(\delta\kappa_{yz}^a\right) + N_{xy}^a\left(\delta\varepsilon_{xy}^a\right) + M_{xy}^a\left(\delta\kappa_{xy}^a\right) \Big) dA \tag{4-36}$$

$$\delta U^m = \int_0^a\int_0^b \Big(N_{xx}^m\left(\delta\varepsilon_{xx}^m\right) + M_{xx}^m\left(\delta\kappa_{xx}^m\right) + N_{yy}^m\left(\delta\varepsilon_{yy}^m\right) + M_{yy}^m\left(\delta\kappa_{yy}^m\right) + N_{xz}^m\left(\delta\varepsilon_{xz}^m\right) + M_{xz}^m\left(\delta\kappa_{xz}^m\right) + N_{yz}^m\left(\delta\varepsilon_{yz}^m\right) + M_{yz}^m\left(\delta\kappa_{yz}^m\right) + N_{xy}^m\left(\delta\varepsilon_{xy}^m\right) + M_{xy}^m\left(\delta\kappa_{xy}^m\right) \Big) dA \tag{4-37}$$

where,

$$\left(N_{ij}^a, M_{ij}^a, F_{ij}^a, G_{ij}^a\right) = \int_{-\frac{h_a}{2}}^{\frac{h_a}{2}} \sigma_{ij}^a\left(1, z, f(z), f(z)'\right) dz, \quad (i, j = x, y, z)$$

$$\left(N_{ij}^m, M_{ij}^m, F_{ij}^m, G_{ij}^m\right) = \int_{\frac{h_a}{2}}^{\frac{h_a}{2}+h_m} \sigma_{ij}^f\left(1, z, f(z), f(z)'\right) dz + \int_{-\left(\frac{h_a}{2}+h_m\right)}^{-\frac{h_a}{2}} \sigma_{ij}^m\left(1, z, f(z), f(z)'\right) dz, (i, j = x, y, z) \tag{4-38}$$

The previously referenced equations utilize the superscripts "a" and "m" to signify the graphene-origami-enabled layer and the magnetostrictive facesheets, respectively. The nonlocal correlations related to forces, moments, and stresses are presented as follows [8, 9]:

$$\left(1-\mu^2\nabla^2\right)\begin{Bmatrix}\{N\}\\\{M\}\\\{F\}\end{Bmatrix}=\begin{bmatrix}[A] & [B] & [S]\\ [B] & [D] & [M]\\ [S] & [M] & [O]\end{bmatrix}\begin{Bmatrix}\{\varepsilon\}\\\{\kappa\}\\\{\chi\}\end{Bmatrix}$$

$$\left(1-\mu^2\nabla^2\right)\begin{Bmatrix}G_{yz}\\G_{xz}\end{Bmatrix}=\begin{bmatrix}0 & L_{44}\\ L_{55} & 0\end{bmatrix}\begin{Bmatrix}\S_{yz}\\\S_{xz}\end{Bmatrix} \tag{4-39}$$

where

$$\{N\}=\left\{N_{xx}\quad N_{yy}\quad N_{xy}\right\}^T,\{M\}=\left\{M_{xx}\quad M_{yy}\quad M_{xy}\right\}^T,\{F\}=\left\{F_{xx}\quad F_{yy}\quad F_{xy}\right\}^T$$

$$\{\varepsilon\}=\left\{\varepsilon_{xx}\quad \varepsilon_{yy}\quad \varepsilon_{xy}\right\}^T,\{\kappa\}=\left\{\kappa_{xx}\quad \kappa_{yy}\quad \kappa_{xy}\right\}^T,\{\chi\}=\left\{\chi_{xx}\quad \chi_{yy}\quad \chi_{xy}\right\}^T \tag{4-40}$$

in which,

$$\begin{bmatrix}A_{11} & B_{11} & D_{11} & S_{11} & M_{11} & O_{11} & L_{11}\\ A_{22} & B_{22} & D_{22} & S_{22} & M_{22} & O_{22} & L_{22}\\ A_{12} & B_{12} & D_{12} & S_{12} & M_{12} & O_{12} & L_{12}\\ A_{44} & B_{44} & D_{44} & S_{44} & M_{44} & O_{44} & L_{44}\\ A_{55} & B_{55} & D_{55} & S_{55} & M_{55} & O_{55} & L_{55}\\ A_{66} & B_{66} & D_{66} & S_{66} & M_{66} & O_{66} & L_{66}\\ \end{bmatrix}=\int_{-\frac{ha}{2}}^{\frac{ha}{2}}\begin{bmatrix}1\\ z\\ z^2\\ f(z)\\ f(z)^2\\ f(z)'^2\end{bmatrix}^T\begin{bmatrix}C_{11}^a\\ C_{22}^a\\ C_{12}^a\\ C_{44}^a\\ C_{55}^a\\ C_{66}^a\end{bmatrix}dz$$

$$+\int_{\frac{ha}{2}}^{\frac{h_a}{2}+h_m}\begin{bmatrix}1\\ z\\ z^2\\ f(z)\\ f(z)^2\\ f(z)'^2\end{bmatrix}^T\begin{bmatrix}C_{11}^m\\ C_{22}^m\\ C_{12}^m\\ C_{44}^m\\ C_{55}^m\\ C_{66}^m\end{bmatrix}dz+\int_{-\left(\frac{h_a}{2}+h_m\right)}^{-\frac{ha}{2}}\begin{bmatrix}1\\ z\\ z^2\\ f(z)\\ f(z)^2\\ f(z)'^2\end{bmatrix}^T\begin{bmatrix}C_{11}^m\\ C_{22}^m\\ C_{12}^m\\ C_{44}^m\\ C_{55}^m\\ C_{66}^m\end{bmatrix}dz \tag{4-41}$$

4.13.2 Extrinsic Labor

The external forces in this inquiry can be ascribed to two factors:

- The Winkler and Pasternak Foundation is a charitable organization.
- The weights exerted on the plane are uniformly distributed.

4.13.3 Thermal Gradient

A multitude of substantial interactions effectively delineate work as a distinct and clearly defined physical quantity. Work is defined as the transfer of energy to or from a nanoplate resulting from the application of an external force. The approach under investigation involves a composite nanoplate composed of magnetostrictive materials and graphene-origami structures. The term "work" denotes the energy applied by external forces along the axes of the nanoplate, leading to its displacement and deformation.

The initial version of the exterior work can be articulated as [10, 11]:

$$\delta W = \int_0^a \int_0^b \left[k_w.\delta w_0 - kg\left(\frac{\partial^2 \delta w_0}{\partial x^2} + \frac{\partial^2 \delta w_0}{\partial y^2}\right) - N_{xx}\left(\frac{\partial^2 \delta w_0}{\partial x^2}\right) - \right.$$

$$\left. N_{yy}\left(\frac{\partial^2 \delta w_0}{\partial y^2}\right) - N_{xx}^T\left(\frac{\partial^2 \delta w_0}{\partial x^2}\right) - N_{yy}^T\left(\frac{\partial^2 \delta w_0}{\partial y^2}\right)\right]$$

$$N_{ii}^T = -\sum_{l=1}^{n} \int_{z_k}^{z_{k+1}} \alpha_{ii}\left(T - T_0\right),\ ii = xx, yy \tag{4-42}$$

The parameters k_w and k_P represent the Winkler stiffness and shear layer stiffness, respectively.

The pre-buckling resultant forces are described in Eq. (4–15).

4.13.4 Kinetic Energy

The fundamental formulation of the equation for kinetic energy can be articulated as follows:

$$\delta T=\frac{1}{2}\iint\left[-I_0\left(\frac{\partial^2 u_0}{\partial t^2}\right)\delta u_0-I_1\left(\frac{\partial^3 u_0}{\partial x\partial t^2}\right)\delta w_0-I_3\left(\frac{\partial^2 u_0}{\partial t^2}\right)\delta\phi_x+I_1\left(\frac{\partial^3 w_0}{\partial x\partial t^2}\right)\delta u_0+\right.$$
$$I_2\left(\frac{\partial^4 w_0}{\partial x^2\partial t^2}\right)\delta w_0+I_4\left(\frac{\partial^3 w_0}{\partial x\partial t^2}\right)\delta\phi_x-I_3\left(\frac{\partial^2 \phi_x}{\partial t^2}\right)\delta u_0-I_4\left(\frac{\partial^3 \phi_x}{\partial x\partial t^2}\right)\delta w_0-I_5\left(\frac{\partial^2 \phi_x}{\partial t^2}\right)\delta\phi_x-$$
$$I_0\left(\frac{\partial^2 v_0}{\partial t^2}\right)\delta v_0-I_1\left(\frac{\partial^3 v_0}{\partial y\partial t^2}\right)\delta w_0-I_3\left(\frac{\partial^2 v_0}{\partial t^2}\right)\delta\phi_y+I_1\left(\frac{\partial^3 w_0}{\partial y\partial t^2}\right)\delta v_0+I_2\left(\frac{\partial^4 w_0}{\partial y^2\partial t^2}\right)\delta w_0+$$
$$\left.I_4\left(\frac{\partial^3 w_0}{\partial y\partial t^2}\right)\delta\phi_y-I_3\left(\frac{\partial^2 \phi_y}{\partial t^2}\right)\delta v_0-I_4\left(\frac{\partial^3 \phi_y}{\partial y\partial t^2}\right)\delta w_0-I_5\left(\frac{\partial^2 \phi_y}{\partial t^2}\right)\delta\phi_y-I_0\left(\frac{\partial^2 w_0}{\partial t^2}\right)\delta w_0\right]dA$$

(4-39)

The mass moment of inertia is defined as follows:

$$I_i=\left(\int_{-h_a/2}^{h_a/2}\Upsilon_i\rho_a dz\right)+\left(\int_{-\frac{(h_a+h_m)}{2}}^{\frac{-h_a}{2}}\Upsilon_i\rho_f dz+\int_{\frac{ha}{2}}^{\frac{(h_a+h_m)}{2}}\Upsilon_i\rho_f dz\right),$$
$$\Upsilon_i=\left[1,z,z^2,f(z),zf(z),f(z)^2\right],\qquad i=(1,2,3,4,5)\qquad (4\text{-}40)$$

4.14 Equation of Motion

The Euler-Lagrange equation related to sinusoidal shear deformation includes the initial variation in strain energy, the work done by applied forces, and the kinetic energy, thereby formulating the five equilibrium equations. The stated conclusion is obtained by putting Eqs. (4–36) to (4–40) into Eq. (3–21), yielding:

$$\delta\boldsymbol{u}_0:\ \frac{\partial N_{xx}}{\partial x}+\frac{\partial N_{xy}}{\partial y}-e_{31}h_cK_cC(t)\frac{\partial^2 w_0}{\partial x\partial t}=I_0\left(\frac{\partial^2 u_0}{\partial t^2}\right)+I_3\left(\frac{\partial^2 \phi_x}{\partial t^2}\right)-I_1\left(\frac{\partial^3 w_0}{\partial x\partial t^2}\right)$$

(4-41a)

$$\delta\boldsymbol{v}_0:\ \frac{\partial N_{xx}}{\partial y}+\frac{\partial N_{xy}}{\partial x}-e_{32}h_cK_cC(t)\frac{\partial^2 w_0}{\partial y\partial t}=I_0\left(\frac{\partial^2 v_0}{\partial t^2}\right)+I_3\left(\frac{\partial^2 \phi_y}{\partial t^2}\right)-I_1\left(\frac{\partial^3 w_0}{\partial y\partial t^2}\right)$$

(4–41b)

$$\delta w_0: \frac{\partial^2 M_{xx}}{\partial x^2}+\frac{\partial^2 M_{yy}}{\partial y^2}+2\frac{\partial^2 M_{xy}}{\partial x \partial y}-k_w.w_0+k_g\left(\frac{\partial^2 w_0}{\partial x^2}+\frac{\partial^2 w_0}{\partial y^2}\right)-N_{xx}\left(\frac{\partial^2 w_0}{\partial x^2}\right)-$$

$$N_{yy}\left(\frac{\partial^2 w_0}{\partial y^2}\right)-N_{xx}{}^T\left(\frac{\partial^2 w_0}{\partial x^2}\right)-N_{yy}{}^T\left(\frac{\partial^2 w_0}{\partial y^2}\right)-e_{31}h_c K_c C(t)\frac{\partial^2 U_x}{\partial x \partial t}-$$

$$e_{32}h_c K_c C(t)\frac{\partial^2 V_y}{\partial y \partial t}=I_0\left(\frac{\partial^2 w_0}{\partial t^2}\right)+I_1\left(\frac{\partial^3 u_0}{\partial x \partial t^2}\right)+I_1\left(\frac{\partial^3 v_0}{\partial y \partial t^2}\right)-I_2\left(\frac{\partial^4 w_0}{\partial x^2 \partial t^2}\right)-$$

$$I_2\left(\frac{\partial^4 w_0}{\partial y^2 \partial t^2}\right)+I_4\left(\frac{\partial^3 \phi_x}{\partial x \partial t^2}\right)+I_4\left(\frac{\partial^3 \phi_y}{\partial y \partial t^2}\right)$$

$$\left(\frac{\partial^4 w_0}{\partial y^2 \partial t^2}\right)+I_4\left(\frac{\partial^3 \phi_x}{\partial x \partial t^2}\right)+I_4\left(\frac{\partial^3 \phi_y}{\partial y \partial t^2}\right) \tag{4–41c}$$

$$\delta \phi_x: \frac{\partial F_{xx}}{\partial x}+\frac{\partial F_{xy}}{\partial y}-G_{xz}=I_3\left(\frac{\partial^2 u_0}{\partial t^2}\right)+I_5\left(\frac{\partial^2 \phi_x}{\partial t^2}\right)-I_4\left(\frac{\partial^3 w_0}{\partial x \partial t^2}\right) \tag{4–41d}$$

$$\delta \phi_y: \frac{\partial F_{yy}}{\partial y}+\frac{\partial F_{xy}}{\partial x}-G_{yz}=I_3\left(\frac{\partial^2 v_0}{\partial t^2}\right)+I_5\left(\frac{\partial^2 \phi_y}{\partial t^2}\right)-I_4\left(\frac{\partial^3 w_0}{\partial y \partial t^2}\right) \tag{4–41e}$$

4.15 Precise Solution for the Buckling of a Nanoplate

Compliance of the proposed solutions with the established boundary conditions of the problem is essential in the Galerkin method. To advance, the critical buckling loads will be determined by substituting the specified response, as indicated in Eq. (4–19), into the governing equation. The Galerkin technique is widely employed in the analysis of buckling phenomena and is a reliable approach for effectively solving partial differential equations with significant accuracy, convergence, and computing economy.

By substituting Eq. (4–19) with Eqs. (4–41a)-(4–41e) and isolating the variables $u_0, v_0, w_0, \phi_x, \phi_y$, the natural frequencies can be derived by resolving the resultant equation. To ascertain the natural frequencies, it is essential to set the determinant of the specified equation to zero.

$$\left|[K]-\omega^2[M]+i\omega[C]\right|=0 \tag{4-42}$$

The matrices K, M, and C denote the stiffness, mass, and damping matrices, respectively.

4.15.1 Quantitative Findings and Analysis

This section intends to deliver a comprehensive analysis of the selected findings from the current study, concentrating primarily on an in-depth examination of the influence of various parameters on the non-dimensional critical buckling load. The parameter representing the magnitude of the magnetostrictive layer, referred to as the core layer, is designated as h_a, while h_m indicates the size of the facesheets.

4.15.2 Thorough Analysis

This section seeks to do a comprehensive analysis of the influence of various parameters on the buckling load of the proposed structures.

4.15.2.1 Impact of the Weight Percentage of Graphene-Origami-Enabled Materials

Table 4.7 has been created to assess the impact of graphene weight percentage and hydrogen atom coverage on the buckling load of plates. The table demonstrates a positive association between the weight percentage of graphene and the buckling load, signifying that an increase in the former leads to an elevation in the latter. Moreover, the data presented indicate that an increase in hydrogen atom coverage results in a decrease in the dimensionless buckling stress of the system.

TABLE 4.7

Impact of the Graphene Weight Fraction (a = b = 40nm, ha = 4nm, hm = 0.1nm, T = 300, $\gamma_1 = -1, \gamma_2 = 0$, SSSS)

	$W_{Gr}\,(wt\%)$					
	0	0.5	1	1.5	2	2.5
Hgr = 100%	8.62323	8.11803	8.12022	8.12008	8.11986	8.1197
Hgr = 90%	8.62323	8.11597	8.11893	8.11925	8.11928	8.11927
Hgr = 80%	8.62323	8.11461	8.11773	8.1183	8.11849	8.11857
Hgr = 70%	8.62323	8.11395	8.1168	8.11746	8.11773	8.11786
Hgr = 60%	8.62323	8.11403	8.11626	8.11689	8.11716	8.11731
Hgr = 50%	8.62323	8.11492	8.11622	8.11669	8.11692	8.11704
Hgr = 40%	8.62323	8.11675	8.11681	8.117	8.11712	8.11719
Hgr = 30%	8.62323	8.11988	8.11824	8.11802	8.11796	8.11793
Hgr = 20%	8.62323	8.12524	8.12114	8.12026	8.11989	8.1197
Hgr = 10%	8.62323	8.13701	8.12828	8.12603	8.12502	8.12446

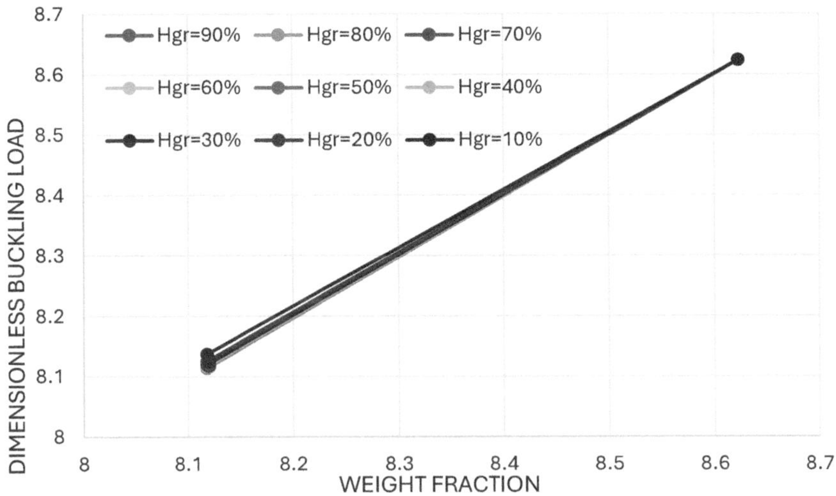

FIGURE 4.7
Influence of weight fraction on non-dimensional buckling loads.

Figure 4.7 has been produced as a persuasive diagram to ascertain the influence of weight fraction on the non-dimensional buckling stress. The figure presents data acquired under the SSSS boundary condition, featuring a nano-square plane with dimensions a = b = 40 nm. It presents an analysis of nine separate values indicating the degree of hydrogen atom coverage.

4.15.2.2 Impact of the Geometry

Table 4.8 illustrates the effect of the magnetostrictive layer's thickness on the buckling load of the continuous system. An increase in the thickness of the magnetostrictive layer correlates with a rise in non-dimensional buckling load, hence enhancing system stability. The table presents the non-dimensional buckling stress for various aspect ratio (a/b) and side-to-thickness ratio (a/h_a) values and illustrates that an increase in the aspect ratio (a/b) results in a reduction of the non-dimensional buckling stress. Furthermore, the non-dimensional buckling stress escalates with an increase in the side-to-thickness ratio (a/h_a).

4.15.2.3 Impact of the Imposed Load

Table 4.9 presents a comparative analysis of the non-dimensional buckling load under various external loads for the boundary conditions of SSSS and SSCC. The examination of the data presented indicates that the non-dimensional buckling loads exhibit increased values under uniaxial

TABLE 4.8

Impact of Geometry ($W_{Gr}(wt\%)$ = 2, Hgr = 100%, T = 320, a = 40nm, $\gamma_1 = -1$, $\gamma_2 = 0, \mu^2 = 1nm^2$)

Magnetostrictive Thickness	B.C	Graphene-Origami Auxetic Thickness	a/b		
	SSSS		1	1/2	1/3
h_m = 0.1 nm		h_a = 5 nm	8.02042	6.29472	7.46676
		h_a = 4 nm	8.02092	6.29511	7.46722
		h_a = 3 nm	8.02178	6.29579	7.46802
h_m = 0.2 nm		h_a = 5 nm	8.02249	6.29634	7.46868
		h_a = 4 nm	8.02358	6.2972	7.4697
		h_a = 3 nm	8.0255	6.29871	7.47149
h_m = 0.3 nm		h_a = 5 nm	8.02472	6.2981	7.47076
		h_a = 4 nm	8.0265	6.29949	7.47242
		h_a = 3 nm	8.02968	6.30199	7.47538
h_m = 0.1 nm	SSCC	h_a = 5 nm	13.5345	18.3176	25.5923
		h_a = 4 nm	13.5353	18.3188	25.5939
		h_a = 3 nm	13.5368	18.3207	25.5967
h_m = 0.2 nm		h_a = 5 nm	13.538	18.3224	25.5989
		h_a = 4 nm	13.5398	18.3249	25.6024
		h_a = 3 nm	13.543	18.3292	25.6085
h_m = 0.3 nm		h_a = 5 nm	13.5417	18.3275	25.606
		h_a = 4 nm	13.5447	18.3315	25.6117
		h_a = 3 nm	13.5501	18.3388	25.6219

TABLE 4.9

Non-Dimensional Buckling Load for Various Nonlocal Parameter Values under SSSS and SSCC Boundary Conditions (a = b = 40nm, ha = 4nm, hm = 0.1nm, T = 320, Wgr = 1.5, Hgr = 100%)

Nonlocal Parameter	(γ_1, γ_2)	SSSS	SSCC
$\mu^2 = 0\ nm^2$	(-1, 0)	8.1201	13.7027
$\mu^2 = 1\ nm^2$		8.02114	13.5357
$\mu^2 = 4\ nm^2$		7.92457	13.3727
$\mu^2 = 9\ nm^2$		7.83029	13.2136
$\mu^2 = 16\ nm^2$		7.73823	13.0583
$\mu^2 = 0\ nm^2$	(-1, -1)	4.06005	6.85133
$\mu^2 = 1 nm^2$		4.01057	6.76784
$\mu^2 = 4\ nm^2$		3.96228	6.68636
$\mu^2 = 9\ nm^2$		3.91515	6.60681
$\mu^2 = 16\ nm^2$		3.86912	6.52913

compression force. The table also includes dimensionless buckling load values corresponding to various non-local Eringen parameter values. An increase in the non-local parameter leads to a reduction in the dimensionless buckling load. The manifestation of this phenomenon is mostly ascribed to the elevation of the non-local parameter, leading to a decrease in the stiffness of the nanoplate. It might be argued that a decrease in the small-scale parameter results in a drop in the system's energy, hence weakening its resilience. Consequently, it can be asserted that the dimensionless buckling loads of the localized system exhibit greater magnitudes than those of the non-local system. The integration of stiffness from the pair stress phenomenon with conventional stiffness results in a cumulative stiffness that surpasses that of the classic model.

Figures 4.8a and 4.8b were generated to analyze the effect of porosity volume on the non-dimensional buckling stress for the SSSS and SSCC boundary conditions. The figures provide clear evidence of a reduction in dimensionless buckling stress with increasing porosity volume. They examine the dimensionless buckling load under uniaxial and biaxial compressive loads.

4.15.2.4 Impact of the Boundary Condition

Table 4.10 provides a thorough analysis of the non-dimensional buckling load for two different boundary conditions: simply supported and simply clamped. The table demonstrates that the buckling load under SSCC boundary conditions surpasses that under SSSS boundary conditions. The improved support encountered under difficult conditions leads to an

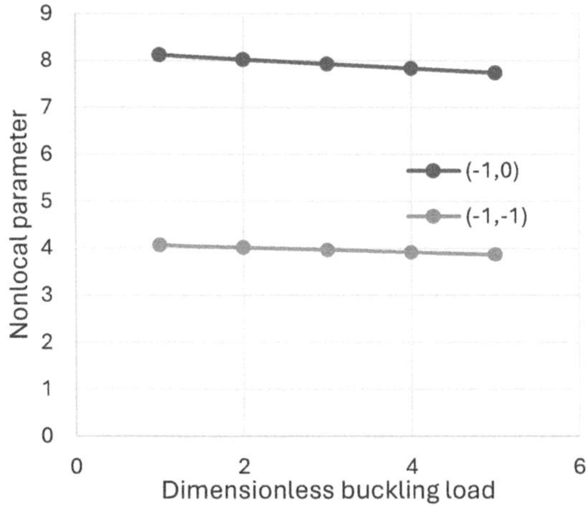

FIGURE 4.8
Influence of the applied load on the non-dimensional buckling load (a) SSSS and (b) SSCC boundary conditions.

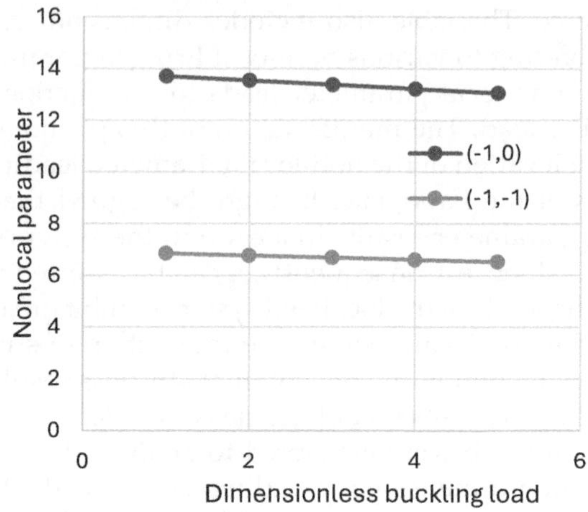

FIGURE 4.8 *(Continued)*
Influence of the applied load on the non-dimensional buckling load (a) SSSS and (b) SSCC boundary conditions.

TABLE 4.10

Comparison of the Non-Dimensional Buckling Load for SSSS and SSCC Boundary Conditions ($\mu^2 = 0nm^2$, a = b = 40nm, hm = 0.1nm, Hgr = 100%, T = 320, Wgr = 1.5)

Ha	(γ_1, γ_2)	SSSS	SSCC
4nm	(-1, 0)	8.1201	13.7027
	(-1, -1)	4.06005	6.85133
3nm	(-1, 0)	8.12105	13.7043
	(-1, -1)	4.06053	6.85214
2nm	(-1, 0)	8.12304	13.7076
	(-1, -1)	4.06152	6.85382
1nm	(-1, 0)	8.12978	13.719
	(-1, -1)	4.06489	6.85951
0.5nm	(-1, 0)	8.14704	13.7481
	(-1, -1)	4.07352	6.87406

increased stiffness in the nanoplate due to the structural modifications it undergoes. The data in the table demonstrate that the non-dimensional buckling stress has an inverse correlation with the thickness of the graphene-origami-enabled layer. This link may be ascribed to the influence of mass and inertia.

TABLE 4.11

Impact of Temperature (a = b = 40nm, ha = 4nm, hm = 0.1nm, Hgr = 100%, $\mu^2 = 0nm^2$, $\gamma_1 = -1, \gamma_2 = 0$)

Wgr	B.C	T (K)						
		300	310	320	330	340	350	360
0	SSSS	8.62323	8.62218	8.62121	8.62029	8.61944	8.61863	8.61787
0.5		8.11803	8.11811	8.11819	8.11826	8.11832	8.11839	8.11844
1		8.12022	8.12025	8.12027	8.12029	8.12031	8.12033	8.12035
1.5		8.12008	8.12009	8.1201	8.12011	8.12012	8.12013	8.12014
2		8.11986	8.11987	8.11988	8.11988	8.11988	8.11989	8.11989
2.5		8.1197	8.1197	8.1197	8.1197	8.1197	8.11971	8.11971
0	SSCC	14.5517	14.5499	14.5483	14.5467	14.5453	14.5439	14.5427
0.5		13.6992	13.6993	13.6994	13.6996	13.6997	13.6998	13.6999
1		13.7029	13.7029	13.703	13.703	13.703	13.7031	13.7031
1.5		13.7026	13.7027	13.7027	13.7027	13.7027	13.7027	13.7027
2		13.7023	13.7023	13.7023	13.7023	13.7023	13.7023	13.7023
2.5		13.702	13.702	13.702	13.702	13.702	13.702	13.702

4.15.2.5 Impact of Temperature

Table 4.11 illustrates the relationship between buckling load and increasing temperature, considering different weight fractions. The table demonstrates a direct correlation between temperature and buckling load, indicating that an increase in temperature results in a decrease in buckling load. The observed phenomena result from minor changes in the lowered stiffness of the GOA composite plate under plane stress, attributable to the decreasing Young's modulus and the growing absolute value of the negative Poisson's ratio.

Furthermore, Figure 4.9 illustrates the effect of temperature on the dimensionless buckling load. The figure demonstrates a slight reduction in the buckling load as the temperature increases.

4.15.3 Final Assessment

This study seeks to examine the buckling characteristics of a composite nano-plate endowed with auxetic properties and incorporated with magnetostrictive facesheets. The investigation employs the higher-order sinusoidal shear deformation plate theory. Additionally, Hamilton's principle is utilized to determine the governing equations, which are then resolved using Galerkin's analytical solution. Several significant findings have been recognized:

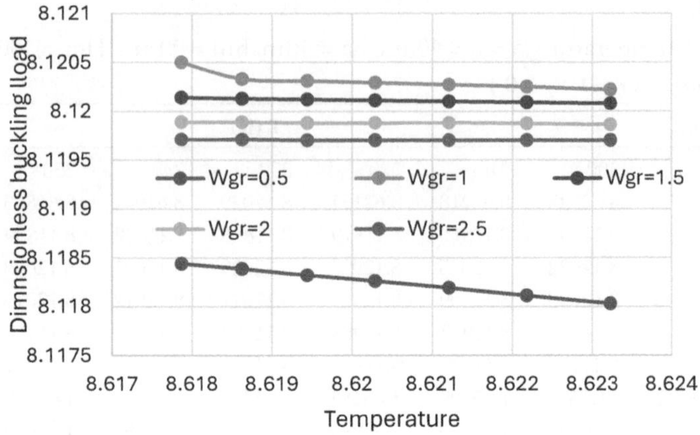

FIGURE 4.9
Dimensionless buckling load as a function of temperature for different levels of velocity feedback gain.

- The buckling load exhibits a favorable correlation with the weight fraction of graphene-origami-enhanced material.
- Augmenting the coverage value of hydrogen atoms results in a decrease in the dimensionless buckling load of the system.
- The confinement of the dimensional buckling load under SSCC boundary conditions is observed to be greater than the dimensionless buckling force under simply supported boundary conditions. The dimensionless buckling load decreases with an increase in the thickness of the magnetostrictive layer.

References

1. Touratier, M., An efficient standard plate theory. *International Journal of Engineering Science*, 1991. 29(8): p. 901–916.
2. Ebrahimi, F., and M.F. Ahari, *Mechanics of Magnetostrictive Materials and Structures*. 2023: CRC Press.
3. Ahari, M.F., and M. Ghadiri, Resonator vibration of a magneto-electro-elastic nano-plate integrated with FGM layer subjected to the nano mass-Spring-damper system and a moving load. *Waves in Random and Complex Media*, 2022: p. 1–39.
4. Ebrahimi, F., and M.F. Ahari, Magnetostriction-assisted active control of the multi-layered nanoplates: effect of the porous functionally graded facesheets on the system's behavior. *Engineering with Computers*, 2023: p. 1–15.

5. Ebrahimi, F., and M. Sepahvand, Wave propagation analysis of cylindrical sandwich shell with auxetic core utilizing first-order shear deformable theory (FSDT). *Mechanics Based Design of Structures and Machines*, 2024. 52(3): p. 1705–1729.
6. Rao, S.S., *Vibration of Continuous Systems*. 2019: John Wiley & Sons.
7. Ebrahimi, F., and A. Dabbagh, Vibration analysis of multi-scale hybrid nanocomposite plates based on a Halpin-Tsai homogenization model. *Composites Part B: Engineering*, 2019. 173: p. 106955.
8. Ebrahimi, F., M.-S. Shafiei, and M.F. Ahari, Vibration analysis of single and multi-walled circular graphene sheets in thermal environment using GDQM. *Waves in Random and Complex Media*, 2022: p. 1–40.
9. Ebrahimi, F., M.-S. Shafiee, and M.F. Ahari, Buckling analysis of single and double-layer annular graphene sheets in thermal environment. *Engineering with Computers*, 2023. 39(1): p. 625–639.
10. Ebrahimi, F., and M.R. Barati, Temperature distribution effects on buckling behavior of smart heterogeneous nanosize plates based on nonlocal four-variable refined plate theory. *International Journal of Smart and Nano Materials*, 2016. 7(3): p. 119–143.
11. Rahimi, Y., et al., Temperature-dependent vibrational behavior of bilayer doubly curved micro-nano liposome shell: simulation of drug delivery mechanism. *Journal of Thermal Stresses*, 2023. 46(11): p. 1199–1226.
12. Mizuji, Z.K., et al., Numerical modeling of a body vessel for dynamic study of a nano cylindrical shell carrying fluid and a moving nanoparticle. *Engineering Analysis with Boundary Elements*, 2023. 152: p. 362–382.

5

Recent Technologies in Active Materials

Active materials are leading the wave of innovation in many fields with the thrust of material science and engineering advancements. Some of the latest technologies and advancements in active materials are given next.

5.1 Smart Textiles

Garments containing electroactive and shape memory material are also being made ready to track health. These materials can be such that they are able to sense physiological properties, that is, temperature and heart rate, and react with property changes based on surrounding conditions. Apart from this, materials changing their properties (thermal insulation or color) upon stimuli (electric, temperature, humidity) are also being brought into use for sportswear and clothing purposes. Smart textiles based on actuator materials are of practical interest, but few types have been commercially exploited. The challenge for researchers has been to bring the concept out of the laboratory by working out how to build these smart materials on an industrial scale and permanently incorporate them into textiles. Smart textiles are considered the next frontline for electronics. Recent developments in advanced technologies have led to the appearance of wearable electronics by fabricating, miniaturizing, and embedding flexible conductive materials into textiles. The combination of textiles and smart materials has contributed to the development of new capabilities in fabrics with the potential to change how athletes, patients, soldiers, first responders, and everyday consumers interact with their clothes and other textile products. Actuating textiles, in particular, have the potential to provide a breakthrough to the area of smart textiles in many ways. The incorporation of actuating materials into textiles is a striking approach as a small change in material anisotropy properties can be converted into significant performance enhancements, due to the densely interconnected structures. Herein, the most recent advances in smart materials based on actuating textiles are reviewed. The use of novel emerging twisted synthetic yarns, conducting polymers, hybrid carbon

DOI: 10.1201/9781003540731-5

nanotube and spandex yarn actuators, as well as most of the cutting-edge polymeric actuators that are deployed as smart textiles, is discussed [1]. Smart textiles are fabrics that have been designed and manufactured to include technologies that provide the wearer with increased functionality. These textiles have numerous potential applications, such as the ability to communicate with other devices, conduct energy, transform into other materials, and protect the wearer from environmental hazards. Research and development toward wearable textile-based personal systems allowing, for example, health monitoring, protection and safety, and a healthy lifestyle has gained strong interest during the past few years. Smart fabrics and interactive textiles' activities include personal health management through integration, validation, and use of smart clothing and other networked mobile devices, as well as projects targeting the full integration of sensors/actuators, energy sources, processing, and communication within the clothes to enable personal applications such as protection/safety, emergency, and healthcare. This writing includes the origin

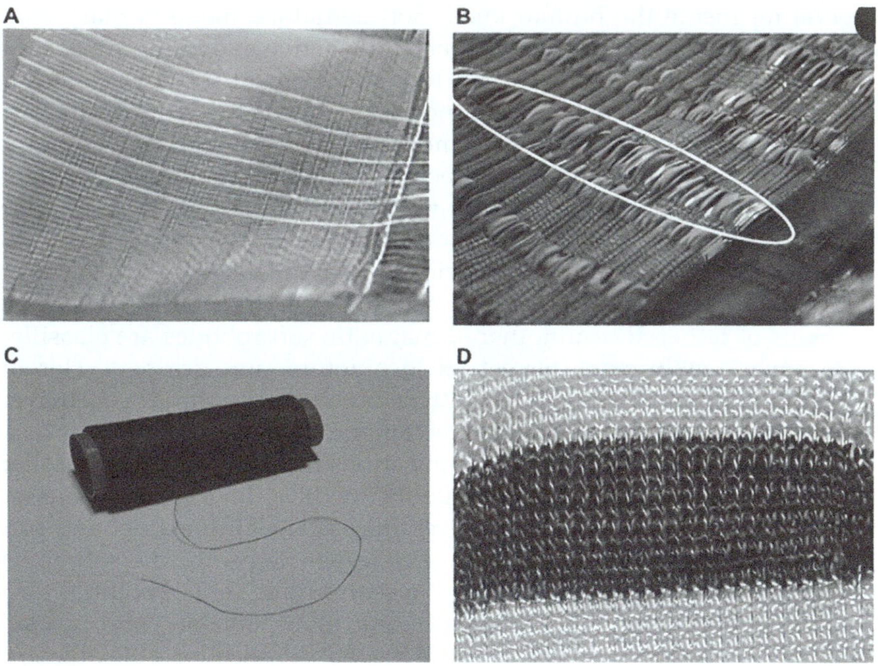

FIGURE 5.1
(A) Cu monofilaments in weave fabric. (B) Custom weave with spacing (marked) allowing yarns to move within the marked region. (C) Bobbin with industrially produced PEDOT-coated yarn. (D) A knitwear structure for respiratory monitoring using CP-coated yarns (black yarn) knitted with regular (white) yarn.

and introduction of smart textiles and integrated wearable electronics for sport wear, industrial purposes, automotive and entertainment applications, healthcare and safety, military, public sectors, and new developments in smart textiles [2].

5.2 Soft Robotics

Emerging innovations in soft robotics utilize shape memory alloys (SMAs) and electroactive polymers (EAPs) to design soft actuators mimicking natural motion, with enhanced dexterity and flexibility of robotic systems. Additionally, biologically inspired soft robots are being designed using active materials to achieve complex movements and environmental interaction. A magnetic field has certain advantages in controlling soft robotics in a contained environment, such as surgical catheters or drug-delivering robots that have no attachment, operating inside the human body. Soft actuators, made of elastomers and gels functionalized with magnetically active materials, are natural choices to drive magnetically controlled motions of soft robots. Recent innovations in soft material technologies, including 3D printing, origami/kirigami, tough hydrogels, mechanical metamaterials, and liquid metal-injected elastomers, offer technological foundations to develop soft actuators and robots with significantly enhanced performance. Herein, an overview of magnetic soft actuators and robots from a materials engineer's perspective is provided. First, the historical background and recent trends of magnetic soft actuators are discussed. Second, the motions of tethered or untethered magnetic soft robotics are classified into aquatic swimmers, terrestrial locomotors, and grippers. Herein, preprogrammed motion under patterned magnetic stimuli is achieved by controlled magnetization of elastomeric materials containing hard magnetic particles. Finally, the applications of magnetically controlled soft robotics in surgical and therapeutic medical devices are discussed [3]. Multiplexed active matrix addressing is critical to extract and process signals from a large number of channels. Recently, the soft active matrix has attracted a great deal of attention due to its potential for a variety of applications such as electronic skins, sensors, and bioelectronics. To realize their use in these various applications, flexible and stretchable formats of the active matrix technologies have emerged. In this review, a variety of reports on the soft active matrix are presented and discussed. The technologies of electronic devices implemented in the modern active matrix are introduced. Strategies to achieve flexible and stretchable active matrix technologies, their advantages and

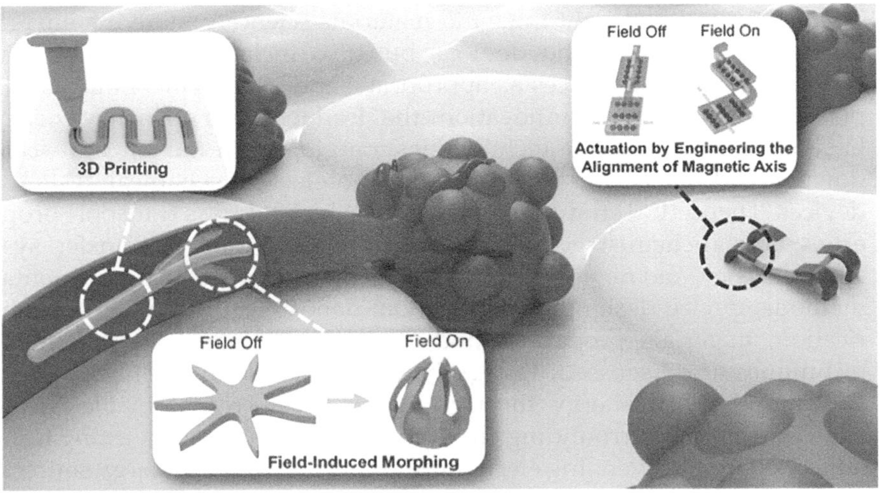

FIGURE 5.2
Diagrammatic representation of terrestrial and swimming robots with magnetic steering nearing tumors [3].

disadvantages, and their applications are briefly described. The review closes with a summary, the associated challenges, and future directions for the development of soft active matrix electronics [4].

5.3 Energy-Harvesting Technologies

New piezoelectric materials and triboelectric nanogenerators are also being explored for vibration and motion harvesting to generate energy with the potential for sustainable power sources for small devices and sensors. Additionally, integration of active material into self-powered sensor systems to tap energy from ambient vibration or heat gradients is growing, reducing battery dependence. Energy harvesting refers to the set of processes by which useful energy is captured from waste, environmental, or mechanical sources and is converted into a usable form. The discipline of energy harvesting is a broad topic that includes established methods and materials such as photovoltaics and thermoelectrics, as well as more recent technologies that convert mechanical energy, magnetic energy, and waste heat to electricity. This article [5] will review various state-of-the-art materials and devices for direct energy conversion and in particular will include multistep energy conversion approaches.

The article will highlight the nano-materials science underlying energy harvesting principles and devices, but also include more traditional bulk processes and devices as appropriate and synergistic. Emphasis is placed on device-design innovations that lead to higher efficiency energy harvesting or conversion technologies, ranging from the cm/mm-scale down to MEMS/NEMS (micro- and nano-electromechanical systems) devices. Theoretical studies are reviewed, which address transport properties, crystal chemistry, thermodynamic analysis, energy transfer, system efficiency, and device operation. New developments in experimental methods, device design and fabrication, nanostructured materials fabrication, materials properties, and device performance measurement techniques are discussed. Energy harvesting is the most effective way to respond to the energy shortage and to produce sustainable power sources from the surrounding environment. The energy harvesting technology enables scavenging electrical energy from wasted energy sources, which always exist everywhere, such as heat, fluids, and vibrations. In particular, piezoelectric energy harvesting, which uses direct energy conversion from vibrations and mechanical deformation to electrical energy, is a promising technique to supply power sources in unattended electronic devices, wireless sensor nodes, micro-electronic devices, etc., since it has higher energy conversion efficiency and a simple structure. Up to now, various technologies, such as advanced materials, micro- and macro-mechanics, and electric circuit design, have been investigated and emerged to improve performance and conversion efficiency of the piezoelectric energy harvesters. In this paper, we focus on recent progress of piezoelectric energy harvesting technologies based on PbZrxTi1-xO3 (PZT) materials, which have the most outstanding piezoelectric properties. The advanced piezoelectric energy harvesting technologies, such as materials, fabrications, unique designs, and properties, are introduced to understand current technical levels and suggest the future directions of piezoelectric energy harvesting [6].

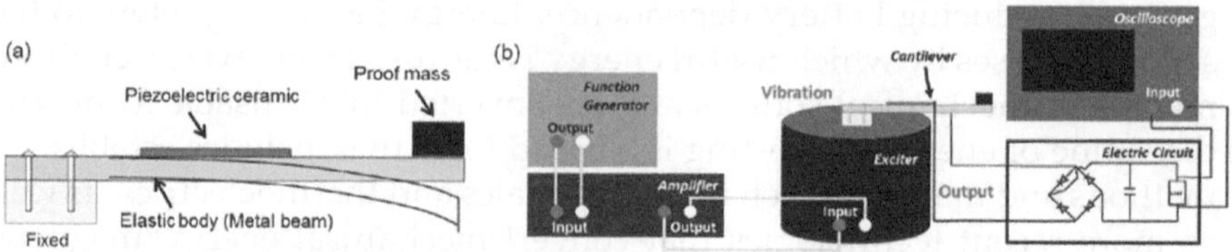

FIGURE 5.3
(a) cantilever-type piezoelectric energy harvester's schematic and (b) a standard experimental configuration for assessing the device [6].

5.4 Shape Memory Alloys and Polymers

Developments in the design and manufacture of SMAs and SMPs are leading to applications in minimally invasive surgery, where devices can change shape once inside the body, enhancing precision and reducing trauma. Research is focused on creating SMPs with programmable shape and function that can be activated by external stimuli, allowing complex designs in a variety of applications. Rapid progress in material science and nanotechnology has led to the development of shape memory alloys (SMA) and shape memory polymers (SMP)-based functional multilayered structures that, due to their capability to achieve properties not feasible by most natural materials, have attracted significant attention from the scientific community. These shape memory materials can sustain large deformations, which can be recovered once the appropriate value of an external stimulus is applied. Moreover, the SMAs and SMPs can be reprogrammed to meet several desired functional properties. As a result, SMAs and SMPs' multilayered structures benefit from the unprecedented physical and material properties such as the shape memory effect, superelasticity, large displacement actuation, changeable mechanical properties, and high energy density. They hold promises in the design of advanced functional micro- and nano-electro-mechanical systems (MEMS/NEMS). In this review, we discuss the recent understanding and progress in the fields of SMAs and SMPs. Particular attention will be given to the existing challenges, critical

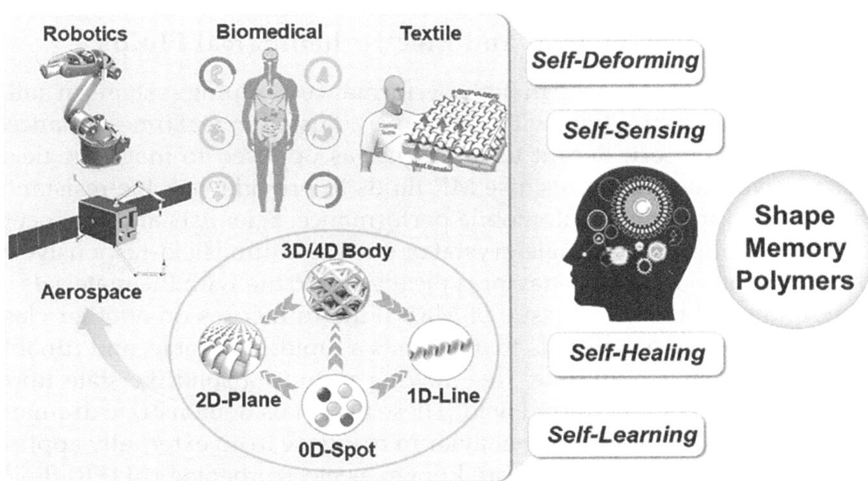

FIGURE 5.4
Schematic of shape memory polymers: actuation techniques, mechanisms, and applications [8].

issues, limitations, and achievements in the preparation and characterization of the SMPs and NiTi-based SMAs thin films, and their heterostructures for MEMS/NEMS applications, including both experimental and computational approaches. Examples of the recent MEMS/NEMS devices utilizing the unique properties of SMAs and SMPs, such as micropumps, microsensors, or tunable metamaterial resonators, are highlighted. In addition, we also introduce the prospective future research directions in the fields of SMAs and SMPs for the nanotechnology applications [7]. Shape memory polymers (SMPs) are one of the primary directions in the development of modern high-tech new materials, which are integrated with sensing, actuation, information processing, and autonomous deformation. Here, multifunctional shape memory polymers are focused and a detailed introduction to the characteristics of self-deformation, self-sensing, self-healing, and self-learning is provided. Integrating with other functional materials to form shape memory polymer composites (SMPC), designing and controlling the material structure and organization at the microscale, thereby achieving more precise and controllable shape memory effects and expanding the potential of material applications. Ultimately, it is shown that SMPs and their composites have a wide range of fascinating applications in the fields of robotics, smart clothing, smart textiles, biomedical devices, and wearable technology. SMPs will thus continue to play a significant role in future deeper exploration [8].

5.5 Magnetorheological and Electrorheological Fluids

MR and ER fluids are used in high-performance damping systems in automobiles and construction, with vibrations adjusted in real time, enhanced safety, and comfort. Recent technologies, as opposed to magnetic field-responsive smart materials, use MR fluids to provide variable-resistance brakes with improved automobile performance. Scientists and engineers are most familiar with single-crystal or polycrystalline field-responsive or "smart" materials with behavior typically occurring with the materials in the solid condition. This issue of MRS Bulletin focuses on another class of field-responsive materials that exhibits a rapid, reversible, and tunable transition from a liquidlike, free-flowing state to a solidlike state upon the application of an external field. These materials demonstrate dramatic changes in their rheological behavior in response to an externally applied electric or magnetic field and are known as electrorheological (ER) fluids or magnetorheological (MR) fluids, respectively. They are often described as Bingham plastics and exhibit a strong field-dependent shear modulus and a yield stress that must be overcome to initiate gross material deformation or flow. Prototypical ER fluids consist of linear dielectric particles (such as silica, titania, and zeolites) dispersed in nonconductive liquids such as silicone oils. Homogeneous liquid-crystalline (LC) polymer-based ER fluids have also been recently reported. MR fluids are based on ferromagnetic or ferrimagnetic, magnetically nonlinear particles (e.g., iron, nickel, cobalt, and ceramic ferrites) dispersed in organic or "aqueous liquids. Unlike ER and MR fluids, ferrofluids (or magnetic fluids), which are stable dispersions of nanosized superparamagnetic particulates (~5–10 nm) of such materials as iron oxide, do not develop a yield stress on application of a magnetic field [9].

5.6 3D Printing of Active Materials

Incorporation of active materials within 3D printing technologies enables the production of complex geometries and structures that respond to environmental stimulants, such as temperature or electric fields. Advances in multi-material 3D printing enable the production of composite structures comprising multiple active materials, with enhanced functionality and performance. Since the pioneering work of Kawata and colleagues in 1997, multi-photon 3D laser printing, also known as direct laser writing, has made significant advancements in a wide range of fields. Moreover, the development and commercialization of photocurable inks for this technique have expanded rapidly. One of the current trends is the transition from static to active printable materials, often referred to as 4D microprinting, which enables a new degree of control in the printed systems. This review focuses on four primary application areas: microrobotics, optics and photonics, microfluidics, and life sciences, highlighting recent progress and the crucial role of active materials, including liquid crystalline elastomers, hydrogels, shape memory polymers, and composites, among others. It also addresses ongoing challenges and provides insights into the future prospects in the different fields [10]. In a research [11], a paradigm of printed active composite materials was realized by directly printing glassy shape memory polymer fibers in an elastomeric matrix. They imbued the active composites with intelligence via a programmed lamina and laminate architecture and a subsequent thermomechanical training process. The initial configuration is created by three-dimensional (3D) printing, and then the programmed action of the shape memory fibers creates time dependence of the configuration – the four-dimensional (4D) aspect.

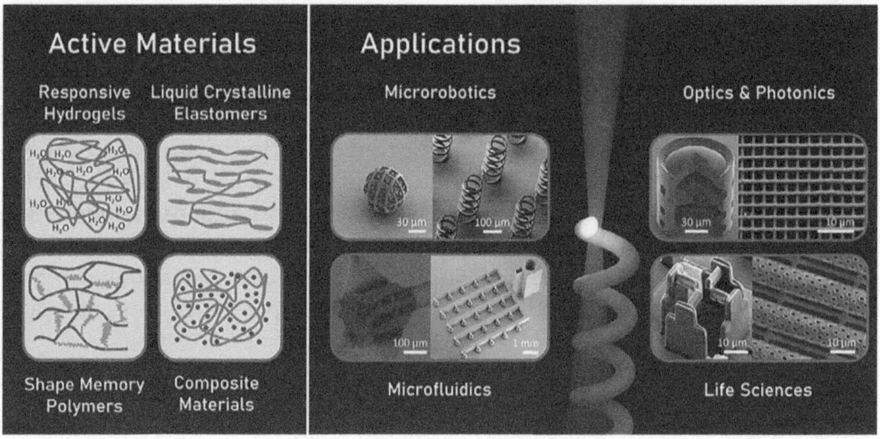

FIGURE 5.5
Overview of the key active materials and four core application areas of multi-photon 3D laser printing [10].

5.7 Artificial Muscles

New technologies focus on the development of EAPs that can be employed as artificial muscles to achieve lightweight and efficient actuation of robotic devices, prostheses, and soft robotics. The combination of different types of active materials (e.g., EAPs with SMAs) creates hybrid systems that benefit from the strengths of each material for higher performance and versatility. The accurate selection of materials and the fine tuning of their properties represent a fundamental aspect in the realization of new active systems able to produce actuating forces, such as artificial muscles. In this regard, exciting opportunities for the design of new advanced systems are offered by materials belonging to the emerging class of functional polymers: exploiting their actuation response, specific devices can be realized. Along this direction, materials showing either shape-memory effect (SME) or shape-change effect (SCE) have been the subject of extensive studies aimed at designing actuators as artificial muscles. Here, we concisely review active polymers in terms of properties and main applications in artificial muscle design. The main aspects related to material properties in both shape-memory polymers (SMPs) and electroactive polymers (EAPs) are reviewed, based on recent scientific literature. SME in thermally activated SMPs is presented by preliminarily providing a definition that encompasses the new theories regarding their fundamental properties. EAPs are briefly presented, describing the working mechanisms and highlighting the main properties and drawbacks, in view of their application as actuators. For both classes of materials, some key examples of effective application in artificial muscles are offered. The potential in polymer architecture design for the fabrication of actively moving systems is described to give a perspective on the main achievements and new research activities [12]. For decades, EAPs received relatively little attention due to their limited actuation capability. However, in the past 15 years a series of Electroactive Polymers (EAPs) materials have emerged that produce a significant shape or size change in response to electrical stimulation. These materials have the closest functional similarity to biological muscles, enabling engineering of novel capabilities that were impossible to do up until recently. Efforts are underway to address the many challenges that are hampering the practical application of these materials, and recent progress has already led to dramatic capability improvements. Various novel mechanisms and devices were demonstrated, including robot fish, catheter steering element, robotic arms, miniature gripper, loudspeaker, active diaphragm, Braille display, and dust-wiper. For developers of future medical devices, these materials offer numerous advantages for their flexibility, fracture toughness, and controllability, as well as low mass and low power requirements. This paper [13] provides a review of the current status, the challenges, and potential near-future applications of these materials.

5.8 Advanced Coatings and Surface Treatments

Active materials are being integrated in coatings that respond to environmental cues, such as self-cleaning surfaces or thermochromic coatings that vary in color by temperature or illumination. In addition, the development of self-healing or adaptive coatings provides greater durability and function. In this review [14], current activities in the development of active biomaterials related to bio-interfacial design, as well as novel concepts for smart polymer systems, are highlighted. For instance, advanced biomedical coatings, including vapor-based polymer coatings for the controlled attachment of biomolecules and temperature-responsive coatings for tissue engineering, are outlined. Moreover, recent trends with the fabrication of "smart," switchable surfaces, such as conformationally switching monolayers, will be reviewed. Specific focus will further be given to multi-functionalized biomaterials, including shape-memory polymers that undergo specific changes in shape in response to stimuli like heat or light. This novel class of intelligent materials highlights a recent trend in biomaterial development, targeting the extension of this concept to multi-functionalization. In particular, the combination of biodegradability and shape-memory effects opens a range of opportunities for biomedical applications, including intelligent sutures or stents.

5.9 Artificial Intelligence in Material Design

The use of AI and machine learning for active material design and optimization is emerging, facilitating rapid discovery and development of new materials with target properties for specific applications. The integration of artificial intelligence (AI) algorithms in materials design is revolutionizing the field of materials engineering, thanks to their power to predict material properties, design de novo materials with enhanced features, and discover new mechanisms beyond intuition. In addition, they can be used to infer complex design principles and identify high-quality candidates more rapidly than trial-and-error experimentation. From this perspective, herein we describe how these tools can enable the acceleration and enrichment of each stage of the discovery cycle of novel materials with optimized properties. We begin by outlining the state-of-the-art AI models in materials design, including machine learning (ML), deep learning, and materials informatics tools. These methodologies enable the extraction of meaningful information from vast amounts of data, enabling researchers to uncover complex correlations and patterns within material properties, structures, and compositions. Next, a comprehensive overview of AI-driven materials design is provided, and its potential future prospects are highlighted. By leveraging such AI algorithms, researchers can efficiently search and analyze databases containing a wide range of material properties, enabling the identification of promising candidates for specific applications. This capability has profound implications across various industries, from drug development to energy storage, where material performance is crucial. Ultimately, AI-based approaches are poised to revolutionize our understanding and design of materials, ushering in a new era of accelerated innovation and advancement [15]. Artificial intelligence (AI) has received widespread attention over the past few decades due to its potential to increase automation and accelerate productivity. In recent years, a large amount of training data, improved computing power, and advanced deep learning algorithms have been conducive to the wide application of AI, including material research. The traditional trial-and-error method is inefficient and time-consuming for studying materials. Therefore, AI, especially machine learning, can accelerate the process by learning rules from

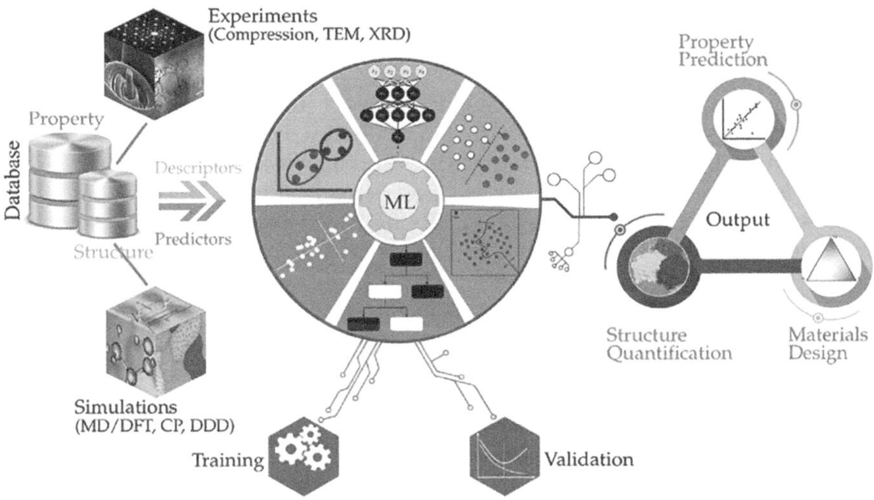

FIGURE 5.6
Schematic diagram of the materials design process with AI, consisting of three main components: a material dataset, machine learning models that have the ability to learn and reason representations for a specific task based on the provided dataset, and output offering optimized and enhanced material properties for the creation of advanced materials [15].

datasets and building models to predict. This is completely different from computational chemistry, where a computer is only a calculator, using hard-coded formulas provided by human experts. Herein, the application of AI in material innovation is reviewed, including material design, performance prediction, and synthesis. Finally, the future development direction of AI is expounded from both algorithm and infrastructure aspects [16].

5.10 Conclusion

Recent advances in active materials are driving innovation across numerous industries, enhancing the performance of smart technologies, robots, biomedical devices, and energy systems. The introduction of new materials, production methods, and design techniques is allowing for the development of more responsive, adaptive, and efficient systems for an extensive array of applications. As research continues, we can expect even more breakthroughs in active materials technology.

References

1. Kongahage, D., and J. Foroughi, Actuator materials: review on recent advances and future outlook for smart textiles. *Fibers*, 2019. 7(3): p. 21.
2. Syduzzaman, M., et al., Smart textiles and nano-technology: a general overview. *Journal of Textile Science and Engineering*, 2015. 5(1): p. 1–7.
3. Chung, H.J., A.M. Parsons, and L. Zheng, Magnetically controlled soft robotics utilizing elastomers and gels in actuation: a review. *Advanced Intelligent Systems*, 2021. 3(3): p. 2000186.
4. Shim, H., et al., Recent advances in materials and device technologies for soft active matrix electronics. *Journal of Materials Chemistry C*, 2020. 8(31): p. 10719–10731.
5. Radousky, H.B., and H. Liang, Energy harvesting: an integrated view of materials, devices and applications. *Nanotechnology*, 2012. 23(50): p. 502001.
6. Kang, M.-G., et al. Recent progress on PZT based piezoelectric energy harvesting technologies, in *Actuators*. 2016: MDPI.
7. Stachiv, I., E. Alarcon, and M. Lamac, Shape memory alloys and polymers for MEMS/NEMS applications: review on recent findings and challenges in design, preparation, and characterization. *Metals*, 2021. 11(3): p. 415.
8. Luo, L., et al., Recent advances in shape memory polymers: multifunctional materials, multiscale structures, and applications. *Advanced Functional Materials*, 2024. 34(14): p. 2312036.
9. Phule, P.P., and J.M. Ginder, The materials science of field-responsive fluids. *MRs Bulletin*, 1998. 23(8): p. 19–22.
10. Mainik, P., C.A. Spiegel, and E. Blasco, Recent advances in multi-photon 3D laser printing: active materials and applications. *Advanced Materials*, 2024. 36(11): p. 2310100.
11. Ge, Q., H.J. Qi, and M.L. Dunn, Active materials by four-dimension printing. *Applied Physics Letters*, 2013. 103(13).
12. Ariano, P., et al., Polymeric materials as artificial muscles: an overview. *Journal of Applied Biomaterials & Functional Materials*, 2015. 13(1): p. 1–9.
13. Bar-Cohen, Y. Electroactive polymers as artificial muscles: capabilities, potentials and challenges, in *Robotics 2000*. 2000: Robotics.
14. Yoshida, M., et al., From advanced biomedical coatings to multi-functionalized biomaterials. *Journal of Macromolecular Science, Part C: Polymer Reviews*, 2006. 46(4): p. 347–375.
15. Badini, S., S. Regondi, and R. Pugliese, Unleashing the power of artificial intelligence in materials design. *Materials*, 2023. 16(17): p. 5927.
16. Sha, W., et al., Artificial intelligence to power the future of materials science and engineering. *Advanced Intelligent Systems*, 2020. 2(4): p. 1900143.

Index

C

D

E

F

T